MIND-BLOWING
MODULAR
ORIGAMI

THE ART OF POLYHEDRAL PAPER FOLDING

BYRIAH LOPER

TUTTLE Publishing

Tokyo | Rutland, Vermont | Singapore

CONTENTS

TRIAKIS
MODULE
16

VORTEX
MODULE
18

EXCALIBUR
KUSUDAMA
20

EXCELSIOR
KUSUDAMA
22

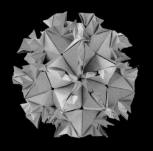

**BOREALIS
KUSUDAMA
24**

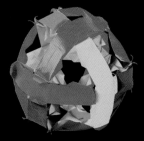

**CURLED
SPHERE
27**

**SATURN
CUBE
30**

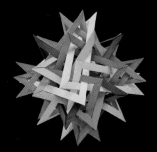

**16
TRIANGLES
36**

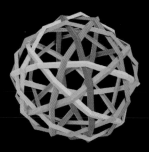

**COSMOS
41**

**NEBULA
46**

**ATMOSPHERE
51**

**K5
56**

**GALAXY
59**

**INTERSTELLAR
63**

**DARK
MATTER
68**

**DARK
ENERGY
74**

**AURORA
79**

**EVENT
HORIZON
84**

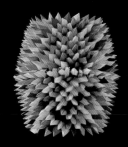

**THE ALPHABET
89**

PREFACE

ORIGAMI, the Asian art of paper folding, has long been seen as an expressive, representative art focused on modeling real-life objects in paper.

From simple airplanes and frogs to highly detailed, lifelike insects and dragons, origami has expanded greatly in the past half-century. New techniques have led to greater complexity, detail, and technical design. During that time, the boundaries of origami have expanded into new concepts, including modular origami.

MODULAR ORIGAMI has most of the same rules as traditional origami, but comprises the folding of multiple pieces of paper into individual units or modules, which are then assembled together to make a single model, generally without the use of any adhesives. Each unit is held to the others strictly by the tension of the paper, and the resulting models cannot be replicated with just a single piece of paper. Modular origami is a wonderful testimony to the remarkable hidden strength of paper, which

is often regarded as being flimsy. As if by magic, hundreds of pieces of this seemingly weak material can be transformed through folding in such a way as to create magnificent three-dimensional structures without the use of any adhesives or cutting. The resulting forms can be surprisingly strong, lasting for years.

Like traditional representative designs, modular origami began very simply. In the 1970s, Mitsunobu Sonobe's development of the Sonobe unit expanded origami to new horizons. This simple universal unit, capable of being assembled into a number of basic geometric polyhedra, paved the way for numerous variations and new designs over the next decades. Tomoko Fuse, Tom Hull, Robert Neale, Miyuki Kawamura, Daniel

Kwan and many others designed increasingly sophisticated constructions; the complexity, innovation and artistry of modular forms has continued to grow. There are thousands of models in existence now, with new designs appearing regularly.

Modular origami isn't limited to polyhedra—flat stars, coasters, tessellated quilts and even representative models have been made with more than one piece of paper—but it was the polyhedral modulars that first caught my attention when I began folding seven years ago. From starkly geometric models that are hardly

more than paper representations of polyhedra to embellished floral spheres so ornate that they seem to have no mathematical basis whatsoever, the symmetry and form that they held attracted me. The repetition of folding the individual units may be tedious, but I have always held that the satisfaction found in the assembly process makes the labor worthwhile. I might never have pursued modular origami beyond a minor hobby, however, had I not discovered the larger origami community online, and with it, a particular type of modular generally referred to as a *Wire Frame*. This category, which includes the most complex and exciting models in modular origami, gave me a renewed interest in this art that has remained with me ever since, and as such, is the focus of this book.

Though I will define it in greater detail later in the book, the term "Wire Frame" describes any intersecting, interlocking or interwoven origami compound, or any model based on the basic "edge unit" which is nearly always used to form the edges of a polyhedron. With these models, the folding of the individual units is invariably much easier than the assembly process. Once I discovered Wire Frames, I folded all that I could find and then started designing my own. Since then, I have designed a great number of these models, and have often been asked to explain how they were made. While several others have diagrammed Wire Frames before, there is no volume devoted primarily to the explanation of these models. So when the opportunity arose to write a book, I took

it as an occasion to help share and explain these geometric origami constructions in a way that could be most easily understood.

While a basic understanding of geometry is helpful, very little knowledge is necessary to begin. With knowledge of the different types of polyhedra and polygons, along with a little experience, you should be able to create all the models in this book, and perhaps even some variations of your own. I selected seven simpler, more embellished modular creations to serve as a basis for the twelve more complicated Wire Frames that follow. If you aren't familiar with modular origami, I suggest starting with the more basic models and getting practice with them until you have gained enough confidence to try the more complicated constructions.

I enjoy designing Wire Frames, and consider them to be the pinnacle of technical modular evolution. With extremely simple units, some remarkably difficult and complicated structures can be made with only paper and the use of your hands. The weaving process is always exciting and challenging. The finished work is a visually stimulating piece that requires more than a passing glance to appreciate. Wire Frame models are akin to geometric puzzles that you must create and then solve using only folded pieces of paper. Like the more complicated representative models, Wire Frames may take more time to assemble than some regular decorative modulars, but the extra effort is apparent in the result.

I have folded and designed all kinds of origami, but have always returned to modular origami, as it is my favorite style. I hope that this book gives you a greater appreciation for the stark geometric beauty and dynamism of modular origami.

MATERIALS

YOU SHOULD KEEP a variety of materials handy for making the models in this book.

Paper is, obviously, the most necessary item to procure. Nearly any paper can be used for origami. I have used many different types, but some are better suited for modular origami than others. The following are several of my favorites. Paper thickness is generally weighed in grams per square meter (gsm). Generally, modulars are best folded with a middle- or heavier-weight paper between 70 and 130 gsm.

COPY PAPER is cheap, colorful, very easy to find, and excellent for Wire Frame modulars. Go for a 24-lb weight.

KAMI, the standard origami paper, is excellent for all of the decorative modulars in this book. While a bit flimsy, it can be used for most Wire Frames.

MEMO PAPER only comes in ~3.5" squares, so it is not generally useful for Wire Frames. However, it is cheap, colorful, and (without a sticky strip), it works very well for decorative models.

TANT is a higher-quality paper that works well for most of the decorative modulars and Wire Frames.

SKYTONE is a higher-quality parchment paper that bears a visual resemblance to Elephant Hide paper, but comes in a greater variety of colors. It is much thinner than Elephant Hide, and isn't as strong.

STARDREAM is a higher-quality paper that is excellent for Wire Frames. It can work for decorative modulars, but its thickness makes it less suitable for some of them.

ELEPHANT HIDE is a very high-quality paper that generally comes in muted colors. It is tremendously strong and thick, but creases superbly. It's excellent for all Wire Frames, and good for most decorative modulars as well.

A bone folder is a piece of bone, plastic or wood used to fold creases strongly. While not necessary, it can help make sharp creases on the center vertices of Wire Frame struts.

Scissors or a paper cutter will be needed to cut the rectangular paper used for Wire Frame struts.

A pencil and ruler will be needed to measure and lightly mark the specifically proportioned rectangles of paper to be cut for most Wire Frame modular units.

Frame-holders can be used to hold the Wire Frame elements, which can be a challenge to keep in place during construction. While not required, frame-holders will give your model added stability until it is complete, at which point they should be removed. Metal wire (22-gauge floral wire) is my frame-holder of choice, but it's not the only possibility. Feel free to experiment.

A protractor is not required, but is useful for checking the angles of the unit's pockets and other angular details.

A calculator is useful for converting the paper proportions for the Wire Frames quickly and easily.

TIPS AND TECHNIQUES

WHILE MANY modular origami projects can indeed be challenging, keeping a few important things in mind ahead of time will make everything go easier.

Before you begin, it is important to be prepared. You should have all the paper that you will need for the project, as well as any tools you might need. Be careful to choose an appropriate size for the starting papers, as they will determine the size of your final model. This is especially true for the Wire Frames—it is very natural for beginners to want to expand the size of the units, but if you aren't careful, you can easily end up with a model several feet in diameter. Conversely, you might decide to make a model very

small, which would then make it quite difficult to fold. Scale the paper proportions up or down to determine an appropriate size.

Paper proportions are the length-to-width ratio of the paper for Wire Frames. In some cases, you may want to change the dimensions listed. Fortunately, this is easy. You just have to convert them to their original 1:X ratio (if necessary), and then multiply both numbers by the desired width. Here's an example: if the paper

proportions are listed as 1.25:5, and you want the width of each unit to be decreased to .875 in order to make a smaller model, you first convert the 1.25:5 back to a 1:X ratio. To do this, divide both numbers by the width. 1.25/1.25=1, and 5/1.25=4, so the proportions have been converted to 1:4. You must now multiply both numbers by the new width to get the final proportion: 1x.875=.875, and 4x.875=3.5. Therefore, your final proportions are .875:3.5. These rectangles will have the same height-to-width ratio as the proportions listed in the directions, only scaled down. The same procedure can be used to increase the proportions; simply reverse the process to increase the proportions.

Once all of the preparations are made, cutting and folding the units is

A batch of units for The Alphabet (page 89), just begun.

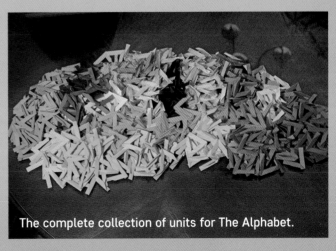
The complete collection of units for The Alphabet.

generally very straightforward, especially for the Wire Frames. However, if you have difficulty with a step, look ahead to the next step to see the result of the fold. The folding of the units can be tedious, but you can fold small "batches" of them and then assemble them later. And the more effort you put into the units, the more rewarding the finished model will be.

The Wire Frame units themselves are surprisingly simple, as they have a standard design pattern. The nature of their design is somewhat redundant, to the point that after having folded several different models, you will probably be able to infer an approximate folding sequence even before reading the diagrams. This will make variations and new concepts easier for you to explore on your own. Probably the most important thing to keep in mind when folding the units, aside from the proportions and pocket angles, is the dihedral angle of each unit. This is the interior angle between the two halves of a completed edge unit. This angle will determine how the

units interact with each other: if it is near 180 degrees, the unit will be close to flat; if it is near 0 degrees, the unit will be narrow, and the two halves will be pressed against each other. Unless a special effect is desired, the optimal angle is around 90 degrees. (It is possible for edge units to have a dihedral angle greater than 180 degrees, but that is a subject for another volume.)

The assembly of the units is usually the most difficult—and the most exciting—part of making modular origami. Various types of locks may be used the hold the units together, but the standard method involves sliding a tab of paper into a pocket. Different paper types each have their own pros and cons in assembly. Thinner paper has the advantage of being more flexible, and will have fewer gaps where the units come together. It is useful in assembly where mobility is limited. Thicker paper is stronger, and is less likely to crumple, bend or rip during the assembly. It can make the completed model more rigid as well. The last units in a modular will be the most

difficult to assemble—especially the solid, ball-like models that offer no way to manipulate the paper into place from underneath once they are near completion. Be patient and deliberate as you slowly ease the units into place.

For Wire Frames, it is important to know where to place any frame-holding pieces to ease assembly. The areas of each unit that press against other units in the assembled model, and which hold the model together, are referred to as the limiting factors.

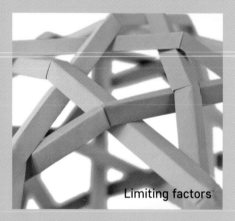

Limiting factors

These determine the proportions of the starting paper. If frame holders are used during assembly, it is important to put them in a place where they will maintain the model's stability, but not interfere with the construction process. The limiting factors are usually the best areas to place any wire, string, etc. that you are planning to use as frame holders.

In addition, the weaving pattern has to be taken into account when assembling the units of the Wire Frames. Getting all of the units woven around each other in the proper pattern so that the model is symmetrical on all sides is a fun puzzle to figure out. To start, having an understanding of basic geometry, especially polyhedra, is absolutely critical to understand the weaving of a Wire Frame. This is because the weaving pattern for any given model will almost always follow the symmetry of a regular

"Neighborhoods" of weaving patterns

polyhedron, most likely one of the Platonic or Archimedean solids (the last model in the book represents the exception to this). For the first two models, I have specified a corresponding polyhedron, but I've left this information to be inferred in later projects. Once you identify what polyhedron symmetry any given model is based on, you can determine which of several methods to use to carry out the weaving and completing the assembly.

The first important key to understanding how Wire Frames are assembled is through their axes. These will be referred to in every Wire-Frame assembly diagram in this book. Strictly speaking, an axis is basically a line around which a figure can be rotated. The axes here will manifest as woven polygonal shapes that form on the model where each frame goes underneath or over another in a rotating manner. This repeats with several others in a circuitous fashion, and the resulting axes align with certain parts of the basic polyhedron on which the model is based. For example, the three-fold axes of a compound might align with the facial viewpoints of an icosahedron, which would equate to aligning with the vertices of a dodecahedron. They are often used to represent different viewpoints in completed models. They are referred to as an n-fold axis, n being the number of sides on the axis.

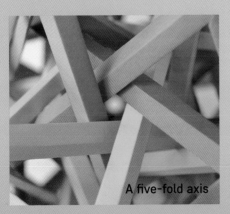

A five-fold axis

Once the concepts of axes and axial weaving are understood, they can be expanded to represent entire "neighborhoods" of the models' weaving pattern. The most common "neighborhoods" in an icosahedral/dodecahedral model, for instance, are the five-fold, three-fold, and two-fold axis views. Each axis, and the surrounding units in its vicinity, represent a "neighborhood" on the surface of the compound; adjacent "neighborhoods" will integrate seamlessly into each other. See the bottom left illustration on the opposite page.

Note that axial weaving alone is not sufficient for more complicated models. These sometimes have double overlapping sections, which can result in illusory axes—areas that have a circuitous whorl in similar frames, but do not exactly represent any polygonal faces of polyhedra.

Another important factor in figuring out the weaving of a complex Wire Frame is identifying if there are any clear relationships between individual frames. One of the most commonly referenced relationships is *in-and-out weaving*. This is an interlocking pattern in which one frame weaves outside of a second frame on one side of the model. On the other side, the frame that was outside now weaves inside the frame that was the inside frame on the other side. Basically, opposite sides of the model are mirror images of one another. See the bottom left illustration on the following page.

Another commonly referenced pattern is *envelope weaving*, where one complete frame is entirely inside of another frame, but entirely outside of a third frame. One example of this is the famous Borromean weaving pattern, where three links are held together through weaving, but any two frames

are not interlocked. See the top right illustration on the following page.

When folding Wire Frames, your initial impulse may be to use the pictures of the assembly in the text to exactly follow the pattern. This will work to a limited degree, but in complex constructions, your view will be obstructed by other parts of the model. In these cases you will have to instead focus on understanding the pattern of each axial area so as to intuit obscured areas based on geometrical patterns, rather than visual images.

Knowing how to weave a complex model is only useful if you are able to physically assemble the units. I have used three different assembly methods; the most practical one will depend on the model you are attempting. The most commonly used method, which was, up until the last few years, the only practiced method, is *frame-at-a-time weaving*. Essentially, any Wire Frame compound is composed of a certain number of identical interlocking polygons, or polyhedra, which are not actually connected to each other, but which interlock around each other in a symmetrical pattern to hold together. In the frame-at-a-time method, you simply assemble one complete frame around another, then add another to the first two, and another, and so on, until the model is completed. This method dates back to the first Wire Frames, including Tom Hull's FIT.

The second method, tailored for the assembly of complex models where most of the units are near the outside surface of the completed piece, is referred to as *bottom-up weaving*. Pieces consisting of perhaps five or ten units of all frames are added simultaneously, so that all frames are assembled with the same progress. The model thus becomes fully completed on the bottom, and more pieces are added to all the frames in a sequentially upward fashion until the compound is complete. One of the advantages of this assembly method is that it makes it easier to weave a Wire Frame in the correct pattern, and it is helpful in understanding the weaving. It is also a particularly useful method if you are experimenting with a new compound idea. The first known use of this method was by Daniel Kwan, with the construction of his compound of Six Irregular Dodecahedra.

The third method, referred to as *scaffolding*, is a hybrid of the previous two; it is generally only used for the most complicated models. It is best used for models that are too complex to be woven with just the frame-at-a-time method, but whose units reach too deeply into the center of the model for stable bottom-up weaving. With this method, as many complete frames as possible are assembled, and then the remaining frames are "bottom-up" woven over the existing "scaffolding," which makes the half-assembled model more stable. I had not seen anyone specifically using this method before I tried it.

Of the various assembly techniques listed here, frame-at-a-time weaving will likely be used the most, followed by bottom-up weaving. Scaffolding weaving will be used the least. The photos of the assemblies in this book show the method I would use for each specific model. In the end, however, the methods you use are up to you.

Another subject that often comes up is coloring guidelines. These aren't specifically mentioned in the instructions themselves. The decorative models can have a variety of coloring patterns depending on the type of assembly. The number of colors should be divisible by the total number of units; i.e., a five-color pattern for a thirty-unit model would require six units per color. I will leave it as an exercise to the reader to figure out the assembly order of the colors. Generally, Wire Frame coloring is basic, especially since all of the Wire Frames in this book are woven compounds. Every frame should get its own color for best results. Alternately, all frames can be colored the same for a special effect, but that will make the separation and interaction between the frames less visible.

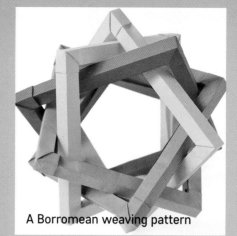

A Borromean weaving pattern

If you don't succeed at your first attempt, simply keep trying. Remember that, like puzzles, these models are supposed to be difficult to figure out. Experience and practice will increase your confidence, and eventually even exceedingly difficult models will seem more reasonable. Follow the path of every frame, and carefully observe how it interacts with every other frame. When the assembly looks impossibly complex, break it down into more manageable parts. Understanding complex models is a matter of accumulating experience. In the end, keep in mind that this book is meant to be an introduction to modular origami, not a complete extrapolation. Wire Frames, and modular origami in general, are too complex to explain fully here.

When you do complete a model, check to make sure everything is assembled, and, if it is a Wire Frame, that the weaving is correct. Be careful during this step—it is very easy to miss a small mistake in the weaving process. When you are sure that it is correct, sit back and take a minute to appreciate your work! I think you will find that it is much easier to fold and appreciate these pieces than it is to find a place for them, especially as you begin to make more.

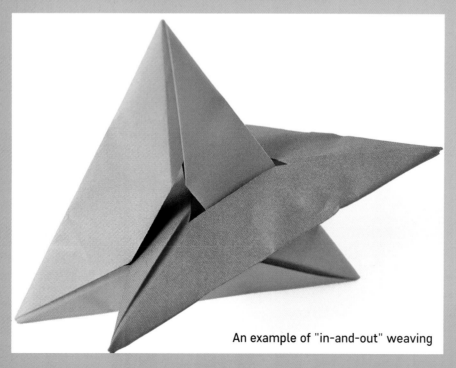

An example of "in-and-out" weaving

ORIGAMI SYMBOLS

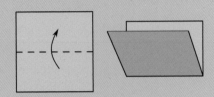

VALLEY FOLD This is the standard fold that creates a "valley" at the base of the fold. It will therefore generally be referred to in diagrams simply as a "fold."

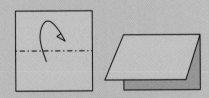

MOUNTAIN FOLD The opposite of a valley fold, where the fold makes a "mountain" at its apex.

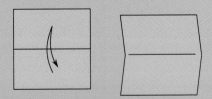

VALLEY FOLD AND UNFOLD A valley fold that is then unfolded to leave a crease.

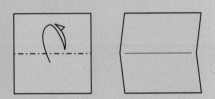

MOUNTAIN FOLD AND UNFOLD A mountain fold that is then unfolded to leave a crease.

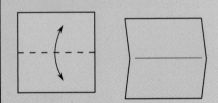

FOLD AND UNFOLD This appears to be the same as the valley fold and unfold, but the arrow is different. This arrow signifies that the fold can be initiated from either side of the paper; the result is the same.

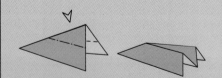

INSIDE-REVERSE FOLD Pressure is applied to the paper along the ridge of the crease, pushing it inside and thus reversing the crease center orientation of the paper.

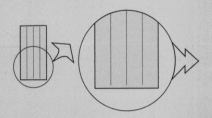

ENLARGE/REDUCE CIRCLES These encircle an area that will be expanded to show small folds with better detail. When the small folds are completed, the diagram will return to its original size, with no circles.

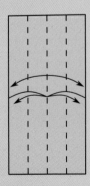

BOOK-AND-CUPBOARD FOLD AND UNFOLD This is a commonly used reference to describe folding a piece of paper in half and unfolding (book), then folding both outer edges in to the center crease, and unfolding (cupboard). This fold is a prerequisite step in all of the Wire Frames in this book.

WATERBOMB BASE

This base isn't used in this book, but is useful for illustrating the techniques that follow.

① Fold diagonals and unfold.

② Mountain fold halves and unfold.

③ Pushing in the center point, collapse the paper, bringing the two edges of the horizontal mountain fold together.

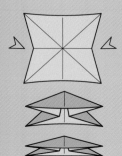

④ Stages of the collapse in progress.

⑤ The completed Waterbomb Base.

OPEN-SINK FOLD

① Start with a Waterbomb Base. Fold the top point to the bottom and unfold.

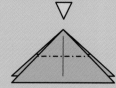

② Open out the top of the Waterbomb Base, making mountain creases around all sides of the fold made in the previous step.

③ Invert the point, pressing on the area indicated by the arrow, and begin to bring the pairs of edges back together.

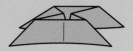

④ In progress.

⑤ The completed Open-Sink fold.

CLOSED-SINK FOLD

① Start with a Waterbomb Base. Fold the top point to the bottom and unfold.

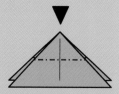

② Keeping the front and back layers together, push the point inside, and begin to invert the paper.

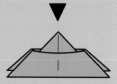

③ In progress. Open the edges slightly if needed.

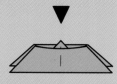

④ In progress.

⑤ In progress. Begin to close the edges back together.

⑥ The completed Closed-Sink fold.

SQUASH FOLD

① Start with a Waterbomb Base. Fold the top layers into the center vertical line and unfold.

② Separate the points' layers on the bottom, and make both sides of the crease mountain folds.

③ Apply pressure to the ridge on the crease, allowing the layers to separate.

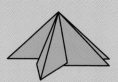

④ In progress.

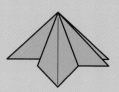

⑤ The completed Squash fold.

SWIVEL FOLD

01. Start with a Waterbomb Base. Fold the top point to the bottom and unfold.

02. Fold the top flap on the left into the center as shown, and unfold.

03. Make a new valley crease between the circled areas through the top layer only. (The paper will not lie flat.)

04. Using the mountain fold from step 2 as a guide, swing the edge over to create a new valley crease.

05. In progress.

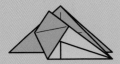

06. The completed Swivel fold.

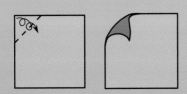

CURL These arrows indicate that the paper should be curled where indicated.

ENLARGE/REDUCE These arrows indicate the next step will be shown larger or smaller.

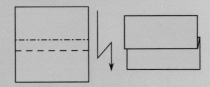

PLEAT SYMBOLS indicate a back-and-forth mountain/valley fold.

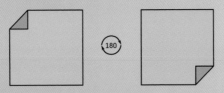

ROTATE This symbol signifies that the paper should be rotated the number of degrees indicated in the direction of the arrows.

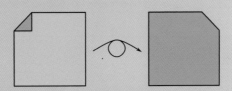

FLIP OVER This symbol means that the paper should be flipped so that the underside is facing up.

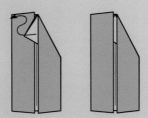

TUCK UNDER This symbol indicates that part of the paper is to be tucked under another layer.

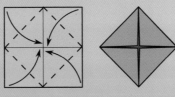

BLINTZ FOLD A simple base where all four corners of a square are folded into the center of the paper.

ANGLE BISECTOR indicates a fold that equally divides an angle in two.

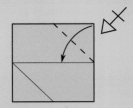

REPEAT SYMBOLS indicate that a step or move should be repeated elsewhere.

PINCH FOLDS don't extend across the entire paper. They are generally used for reference creases.

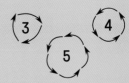

AXIAL SYMBOLS are used to represent axes on Wire Frame modulars. Above are the symbols for three-, four- and five-fold axes, respectively.

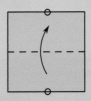

CIRCLES indicate areas of the paper that the fold is meant to join, or areas between which a fold should be made.

PLATONIC SOLIDS

THERE ARE MANY different polyhedra in geometry, but for the purposes of this book, I have only included images of the Platonic Solids.

The Platonic Solids are the five basic regular convex polyhedra comprising a single type of regular polygon faces.

Tetrahedron—six edges, four faces, four vertices.

Hexahedron/cube—twelve edges, six faces, eight vertices.

Octahedron—twelve edges, eight faces, six vertices.

Dodecahedron—thirty edges, twelve faces, twenty vertices.

Icosahedron—thirty edges, twenty faces, twelve vertices.

Applying polyhedra in the assembly process

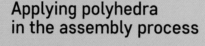
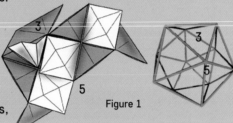

Figure 1

Each origami unit will represent an edge, face, or vertex of a polyhedron; usually an edge. For instance, in Figure 1, above, each unit represents one edge of an icosahedron, as marked by the red lines. Five units will meet at each vertex, three at each face. Identifying what part of a polyhedron the unit represents, as well as what polyhedron is being represented, will make the assembly simpler. While the illustration above is from a Kusudama, this same process applies to Wire Frame modulars, although the interwoven units makes the pattern less obvious.

The connection between polyhedra

Figure 2

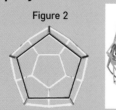

The Kusudamas are not just a warm-up for the folding process; they are to familiarize you with the basic polyhedra behind the more complicated Wire Frames. Once you understand the dodecahedron on the left in Figure 2 (below), the model on the right will be easier to understand. In the first few Wire Frame models in this book, potential underlying polyhedra will be given, but later in the book I will leave it to you to interpret the models' symmetry, and will focus more on axial weaving and overarching assembly techniques. Note also that some models have a more direct resemblance to an Archimedean Solid. Archimedean Solids have all the same conditions as Platonic Solids, except that they allow for the polyhedron to have more than one kind of regular convex polygon. (Because of their symmetry, however, they exclude models with dihedral group symmetry, such as prisms.)

All the models in this book can still be linked back to a Platonic Solid, so I won't include images of the Archimedean Solids group. (If you are interested, these can be found in a variety of mathematical textbooks and online sources.) There are other sets of polyhedra as well, each bound by various parameters, such as the Catalan, Kepler-Poinsot, and Johnson Solids. For this volume, however, a basic understanding of the Platonic Solids and how they can be applied in assembling polyhedral modular origami will be sufficient.

INTRODUCTION TO
DECORATIVE MODULARS

WHILE THIS BOOK is indeed mostly devoted to the serious, mathematically complex, and often difficult Wire Frame modulars, it actually begins with a series of seven simpler decorative modulars.

I like designing and folding these simpler models, because I have always designed modulars from more of an artistic perspective. After all, although mathematics makes appearances in origami everywhere, it is still an art form first and foremost, and in modulars, that can be most obviously seen in the decorative modulars, also known as Kusudamas. It is commonly misunderstood that Kusudamas are not actually origami, but rather are paper craft creations that require the use of string or glue to hold together; compromises that purist origami artists would never allow. While it is true that Kusudamas were first realized in this manner, their definition has evolved over time to include embellished/decorative modular origami where artistic decoration is given a greater emphasis than geometric form. These models are an excellent place to begin if you are just getting started with modular origami as a whole, as they will introduce you into a number of techniques that will be used in the Wire Frame models, but

require less time and patience to master. Generally, the ideal Kusudama is attractive, easy to assemble, and sturdy when completed. I hope you will find the ones I have chosen for this book to match that description. In addition to having similar units, another thing that sets Wire Frames apart from most other modulars is that they have a fixed number of units. Kusudamas and regular geometric modulars can usually be assembled in several fashions, for instance, tetrahedral, octahedral, and icosahedral,

and a different number of units are required for each. This is not the case with Wire Frames, where the underlying symmetry of a model can only be made to align with a single polyhedron, for instance, a dodecahedron. A Wire Frame with dodecahedral symmetry cannot be assembled in an octahedral fashion. It is often quite likely there is an octahedral version, but it would require different paper proportions, possibly different angles, and may not weave together in the same fashion.

TRIAKIS MODULE

THIS MODEL IS very simple in design, but it is one of my favorites. It uses a "wrap-around" lock, rather than the standard tab-and-pocket type. I used to think it looked like a dual polyhedron known as a triakis icosahedron. I have since realized that there actually isn't a close resemblance, but the name stuck. I also realized that the model isn't even mathematically perfect—the angles are a few degrees off, demonstrating that there is more to origami than geometry. This model can be assembled so that the triangles are convex, or, as shown here, concave.

3.5" squares make a ~6" icosahedral assembly. Start colored side up.

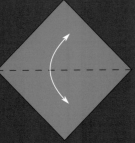

① Fold the paper in half diagonally, and unfold

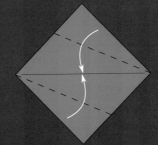

② Fold opposite angle bisectors as shown

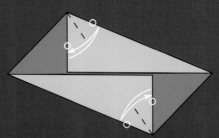

③ Make pinches through the top layers only, and then unfold

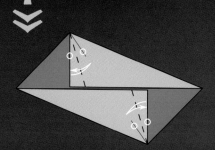

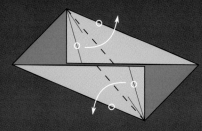

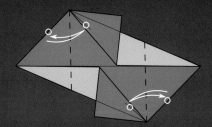

④ Working with the top layer only, fold in the outside edges to the pinches made in the previous step. Unfold.

⑤ Working with the top layer only, fold the crease made in the previous step to the outside edge.

⑥ Fold in the outer edges to the pinch made in Step 3. Unfold.

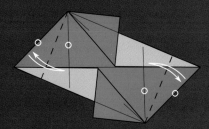

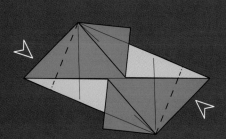

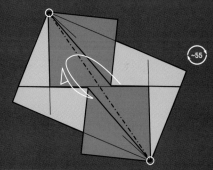

⑦ Fold in the outer edges to the fold made in the previous step. Unfold.

⑧ Inside-reverse fold along the creases made in the previous step.

⑨ Mountain fold the unit in half. Unfold.

⑩ The completed unit.

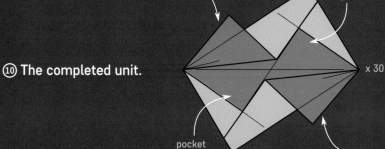

tab

pocket

pocket

tab

x 30

ASSEMBLY

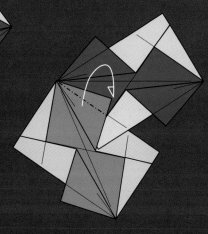

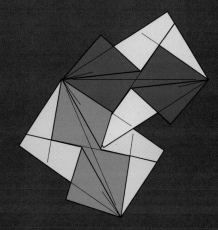

① Slide the tab over the top of the adjacent unit, while sliding the blunt edge of the adjacent unit between the top and bottom layers of the first unit.

② Mountain fold along the line made in Step 4 (above), and tuck into the pocket below.

③ The result. Lock the other units in the same fashion.

VORTEX MODULE

THE VORTEX module is one of the simplest and most straightforward in this book. Its locks create a beautiful swirling effect. Its strength and versatility make it well suited for larger assemblies and variations. It can be folded concave or convex, with points composed of three to six units, allowing for the formation of Platonic, Archimedean, and even Johnson solids. As a special note, these units make a lovely snub disphenoid.

3" squares make a ~4.5"-tall dodecahedral assembly. Start white side up.

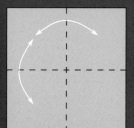

① Fold in half in both directions and unfold.

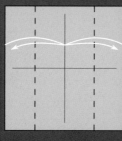

② Cupboard fold and unfold.

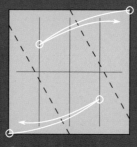

③ Pivoting along the center crease, fold as shown, and then unfold.

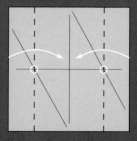

④ Fold in the edges along the circled intersections.

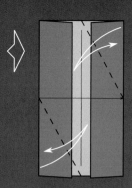

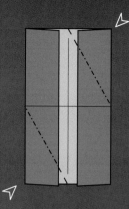

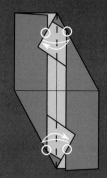

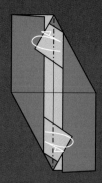

⑤ Refold along the creases made in Step 3; unfold.

⑥ Inside-reverse fold along existing creases.

⑦ Fold through the top layer only; unfold.

⑧ Recrease these folds as mountain folds; unfold.

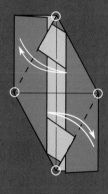

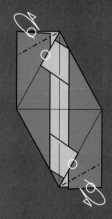

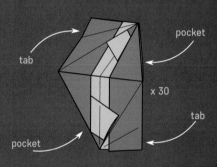

⑨ Fold and unfold the tabs.

⑩ Mountain fold the locks; unfold.

⑪ Mountain fold the unit in half and allow to unfold partially.

⑫ The completed unit.

 ASSEMBLY

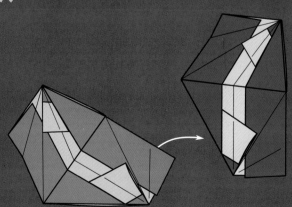

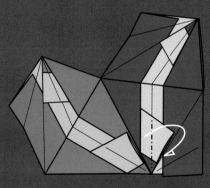

① Slide the tab into the pocket, aligning the creases made in Steps 8 and 10, above.

② Refold the mountain folds from Step 8 through both layers, and tuck underneath to lock the units.

③ Two units connected. Join the others in the

EXCALIBUR KUSUDAMA

THIS IS an example of an embellished star ball. There are hundreds of wonderful variations on this type of model by dozens of different designers. This one was a result of experimentation with more unusual angles, as I think the folding process is as important as the finished result. The locks are very strong and slightly difficult to assemble; getting the last unit in may be something of challenge. The sword-like embellishments that inspired the name lend themselves well to variation, so feel free to experiment with ideas of your own.

6" paper makes a 5.5"-tall model. Begin white side up.

Fold diagonals
d then unfold.

② Fold opposite corners into the center.

③ Mountain fold in half.

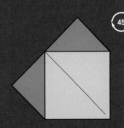

④ The result. Rotate 45 degrees.

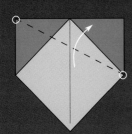

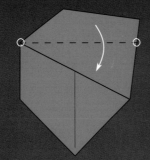

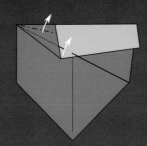

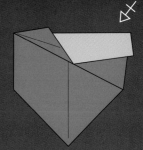

05 Fold up as shown.

06 Fold down to align with the layer behind.

07 Swivel the paper up to align with the layer behind; squash flat.

08 The result. Flip over and repeat Steps 5–7.

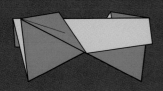

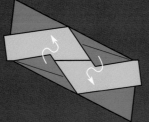

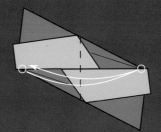

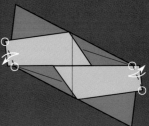

09 The result.

10 Tuck opposite edges underneath the top layer.

11 Fold in half and unfold.

12 Fold and unfold as shown.

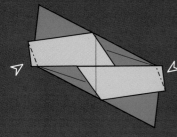

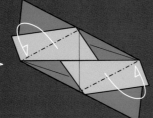

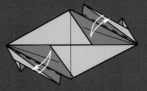

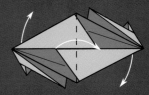

13 Inside-reverse fold along the creases made in the previous step.

14 Mountain fold opposite tabs.

15 Fold the top layer of the tabs down along the edge of the unit, and then unfold.

16 Unfold Step 14. Refold the center crease lightly.

The exterior embellishments can be curled or modified as desired to create different effects.

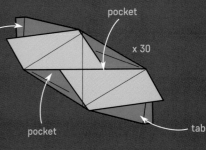

pocket

tab

x 30

pocket

tab

17 The completed unit.

 ASSEMBLY

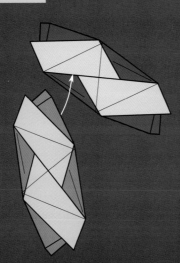

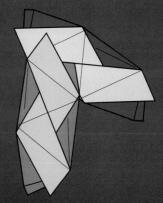

01 To assemble the units, slide the bottom layer of one tab into the pocket of a second unit.

02 The result. Assemble the other units in the same fashion.

EXCELSIOR KUSUDAMA

THIS MODEL is one that I designed very quickly, but refined very slowly. I originally imagined it as a perfectly angled flat dodecahedron, but the locks tended to pull out. Changing the angles to have slightly concave faces fixed the problem sufficiently, and eased the complexity of the resulting units considerably. A paper that holds creases well is a good choice for this model. This model was one that ended up being much more popular than I expected. The name, which means "ever upward," was inspired by an especially productive period of origami design for me.

3" paper makes a 4.5"-tall model. Begin colored side up.

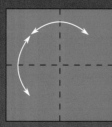

① Fold in half and unfold.

② Cupboard fold and unfold.

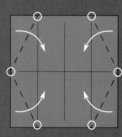

③ Fold the edges in as shown.

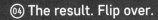

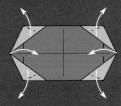

④ The result. Flip over.

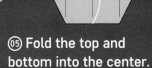

⑤ Fold the top and bottom into the center.

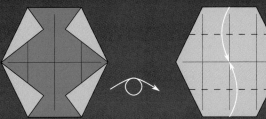

⑥ The result. Flip over.

⑦ Fold and unfold the edges.

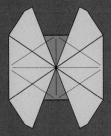

⑧ Refold the corners back in while folding out the sides; squash flat.

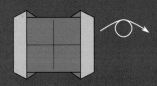

⑨ The result. Flip over.

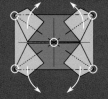

⑩ Fold outward as shown and squash flat.

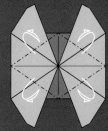

⑪ Mountain fold all four sides; unfold.

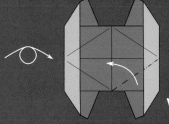

⑫ The result. Flip over.

⑬ Closed-sink fold the bottom right corner in along existing creases.

⑭ In progress. Push in the point where shown.

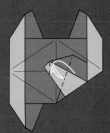

⑮ The result. Mountain fold and unfold as shown. (This will result in a small extra fold at the top as well.)

⑯ Repeat Steps 13–15 on the upper left corner.

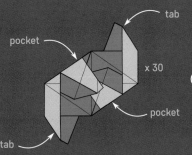

tab

pocket

x 30

pocket

tab

⑰ The completed unit.

ASSEMBLY

① Slide the tab into the pocket.

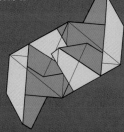

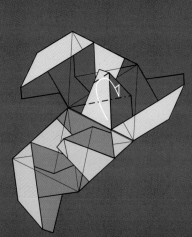

② Mountain fold inside firmly along the line made in Step 14. Again, the tip will be flattened inside in order to fit.

③ The result. Join the other units in the same fashion.

BOREALIS KUSUDAMA

WHILE THIS design, named for its icy-looking exterior flaps, appears to be a variant of the famous Sonobe unit, it does not lock by friction as the Sonobe does. Rather, the decorations lock the model together, similar to the Curled Sphere (page 27). I sometimes use this unit for constructions of more than thirty units, like the sixty-unit stellated rhombic triaconta-hedron assembly shown here. I always take pleasure in cre-ating and folding designs with functional embellishments.

3.5" paper makes a 3.5" model. Begin white side up.

① Book fold in both directions and unfold.

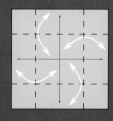

② Cupboard fold in both directions and unfold.

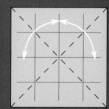

③ Fold diagonals in both directions and unfold.

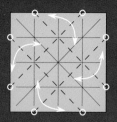

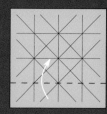

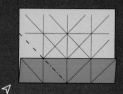

④ Fold as shown, using the circled areas as reference points.

⑤ Fold up the bottom quarter.

⑥ Inside-reverse fold along the existing crease.

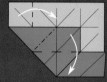

⑦ Fold the top layer over.

⑧ Make a swivel fold along existing creases.

⑨ Fold the top flap over to the left.

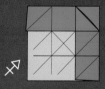

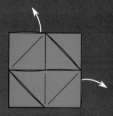

⑩ The result.

⑪ Repeat Steps 5–10 on this side of the paper.

⑫ The result. Pull out some paper from inside and flatten the resulting points.

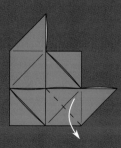

⑬ Swing the bottom flap down.

⑭ Fold in opposite corners towards the center.

⑮ Trisect the top points and partially unfold.

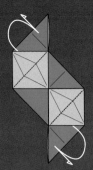

⑯ Mountain fold and unfold as shown.

⑰ Unfold the creases made in Step 14. Then swing the points over and flatten.

⑱ The result. Mountain fold the tabs and unfold partially. Don't fold the top layer.

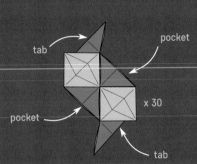

tab

pocket

pocket

x 30

tab

⑲ Fold the unit in half and unfold partially.

⑳ The completed unit.

ASSEMBLY

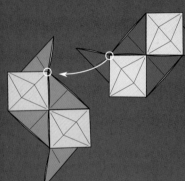

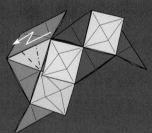

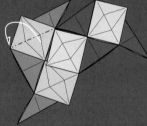

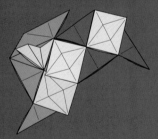

⓵ There are three steps in the assembly process of any two units. First, slide the tab into the pocket, aligning the circled areas.

⓶ Mountain fold along the crease made in Step 14, which should align with the crease made in Step 16 on the other unit.

⓷ Refold the angle trisections through both layers to lock the units.

⓸ The result. Join the other units in the same fashion.

CURLED SPHERE

THIS MODULE is an unusual one, both in its folding pattern and in its result. I have made a variety of models with the curved lock used here, but this one is undoubtedly the most bizarre. The units' svelte look may lead you to think that long rectangles were used, but surprisingly, like all the other Kusudama models in this book, each unit uses a square. Although each piece takes several minutes to fold, the unique result is well worth the effort. The curls not only add a fun embellishment but also function as the key to the locking mechanism.

6" paper makes a ~6"-tall octahedral assembly. Start white side up.

01 Fold diagonals and unfold. Then flip over.

02 Fold opposite corners into the center.

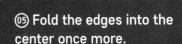

③ **Fold the top layer out to the edges as shown.**

④ **The result. Flip over.**

⑤ **Fold the edges into the center once more.**

⑥ **Make two mountain folds where shown, through all layers, and unfold.**

⑦ **Make valley folds where shown on both sides; unfold.**

⑧ **Make several swivel folds through the top layer only.**

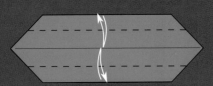

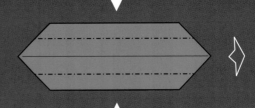

⑨ **In progress.**

⑩ **The result. Repeat Steps 8–9 on the other three sides of the model.**

⑪ **The result. Pull out some paper from the edges.**

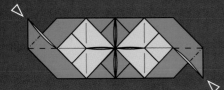

⑫ **Squash the resulting points flat.**

⑬ **Fold the top layer out on both sides.**

⑭ **The result. Flip over.**

⑮ **Fold in all layers except for the center points to the center and unfold.**

⑯ **Closed-sink fold both sides in along the creases made in the previous step. Again, this should not include the center points.**

⑰ **The result. Flip over.**

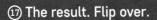

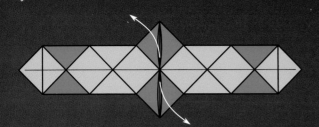

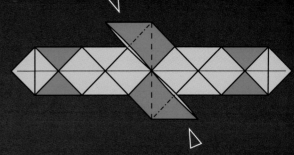

⑱ Pull out some paper from inside; squash flat.

⑲ Squash fold the resulting points.

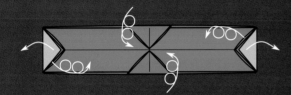

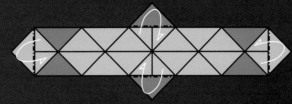

⑳ Mountain fold the top layers.

㉑ Mountain fold all four points through all layers. Flip over.

㉒ Curl up all four points tightly. Don't curl the outer points on the left and right edges. Also, partially fold out the top layers on the left and right points.

㉓ Curl the entire unit up lightly. You will not be able to curl the center.

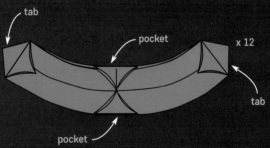

tab

pocket

x 12

tab

pocket

㉔ The result. This is the completed unit.

ASSEMBLY

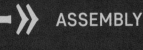

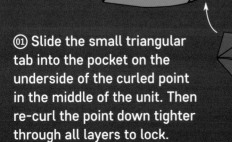

① Slide the small triangular tab into the pocket on the underside of the curled point in the middle of the unit. Then re-curl the point down tighter through all layers to lock.

② The curls made through the whole unit should re-form during the assembly.

SATURN CUBE

MOST ORIGAMI

cubes are composed of either six units—one per each face of a cube; or twelve units—one per each edge of a cube. This model has only four units, each one representing one face and a quarter of the two remaining faces. The units form a ring around the center cube (hence the name). The units are fairly complex, so relatively large paper is recommended. While larger cubic assemblies can be made with this unit, there are thirty steps per unit, so four units might be enough for you!

10" paper makes a ~3"-tall model. Begin color side up.

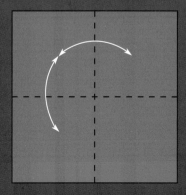

① Book fold in both directions; unfold.

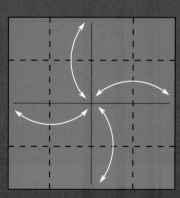

② Cupboard fold in both directions; unfold.

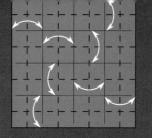 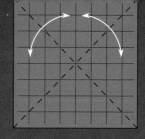 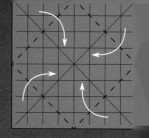

③ Fold eighths and unfold.

④ Fold diagonals and unfold.

⑤ Blintz fold.

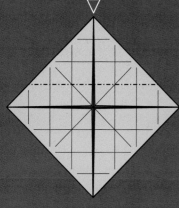 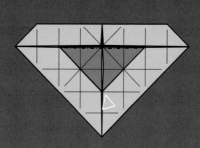 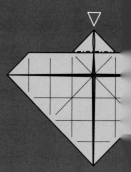

⑦ Open-sink fold to line up with the center.

⑧ Open-sink fold the

⑥ Open-sink fold along the first line above the center.

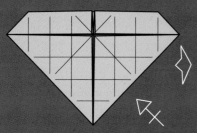 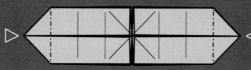 

⑨ The result. Repeat Steps 6–9 on the bottom of the paper.

⑩ The result. Open-sink fold the side points in.

⑪ Fold the top layer o squashing some pape where the arrow is po

 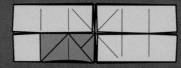 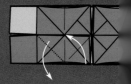

⑫ In progress.

⑬ The result. Repeat on the other three sides.

⑭ Swivel fold as sho

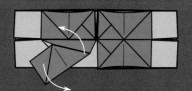  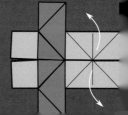

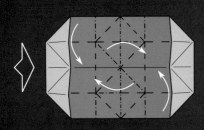  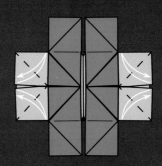

⑱ Collapse along the creases as shown, through the top layer only, and collapse the sides back into the center.

⑲ In progress.

⑳ The result. Fold and unfold the top layer down as shown on all four sides.

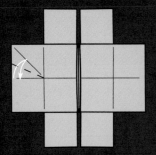 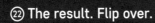 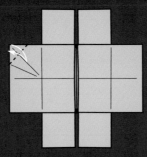 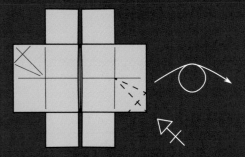

㉑ Squash fold along angle bisectors.

㉒ The result. Flip over.

㉓ Pinch fold as shown.

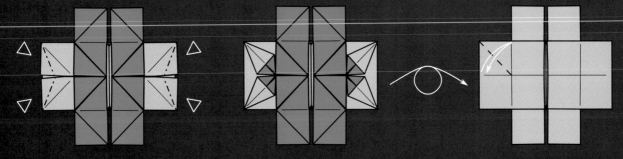

㉔ Pinch fold along an angle bisector.

㉕ Fold down and unfold the edge, using the crease made in the previous step as a guide.

㉖ The result. Repeat Steps 23–25 on the bottom right. Then flip over.

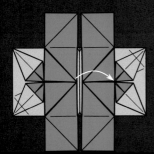 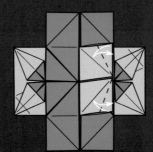 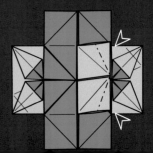

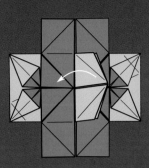

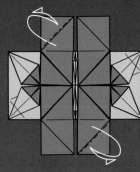

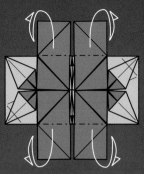

㉚ Return the center point to its vertical orientation.

㉛ Mountain fold and unfold opposite tabs.

㉜ Mountain fold and unfold all four tabs.

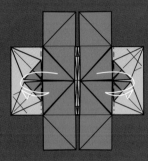

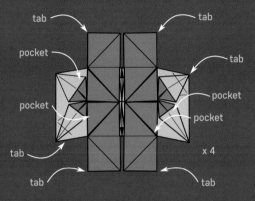

㉝ Mountain fold the right and left sides along existing hinges, and unfold partially.

tab

tab

pocket

tab

pocket

pocket

pocket

tab

tab

tab

tab

x 4

㉞ The completed unit.

ASSEMBLY

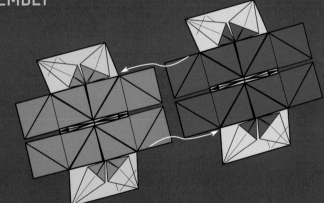

⓵ Slide two opposite tabs underneath the other tabs and into the pocket areas.

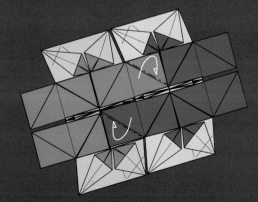

⓶ Mountain fold the triangles from the two remaining tabs into the pockets behind.

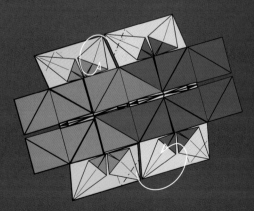

⓷ Mountain fold the top and bottom tabs made in Steps 23–26 into the pockets made in Step 21. At this point the model will be 3D. Assemble the remaining two units in the same fashion.

INTRODUCTION TO
WIRE FRAMES

WIRE FRAMES by themselves are a topic upon which volumes could be written.

As stated earlier, Wire Frames represent a distinct category of modular origami, with a number of differences significant enough to merit being placed in a special subgroup. In theory, any work of modular origami composed of non-embellished "edge units" that represent the edges of a polyhedron could be considered a Wire Frame. As you scan through the Wire Frames in this book, you might be surprised at how similar the units are in design, even when the models are so distinctly different. The typical edge unit has lengthwise book-and-cupboard folds and an angled pocket on either side. This is part of what makes Wire Frame models unique, and in the end is the key determinant as to whether or not a modular qualifies as a Wire Frame. For the purpose of this book, a Wire Frame is defined as any geometric modular origami composed of "edge units" that involve a number of identical individual polygons or polyhedra interlocked around one another to form a symmetrical compound. Each polygon or polyhedron in the compound, which is referred to individually as a "frame," is a separate object that is woven around others, but isn't connected to them. The width-to-length ratio of the edge units determines the pressure each frame exerts on the others. Getting this ratio

correct is essential—too narrow and the frames will not be able to support each other well, allowing the model to sag; but too tight, and the edge units will be forced to bend and wrinkle.

Wire Frames have a rather short history, and only a handful of people have contributed to them. In 1993, Tom Hull got the idea of using a simple unit designed by Francis Ow to create a modular based on the five-tetrahedron compound, and the first Wire Frame was created: Five Intersecting Tetrahedra, or FIT for short. A few years later, Robert Lang proved that many more polygon and polyhedron compounds could be realized. He created a set of fifty-four uniform compounds (which he referred to as polypolyhedra, because there were "many" polyhedra in each construction), and folded five of them. These were bound by a distinct set of rules, and thus were limited. Around 2000 or so, Daniel Kwan demonstrated that if the rules of the polypolyhedra set were abandoned, many new possibilities arose. Over the course of the past decade, he has contributed dozens of new Wire Frames, including the first irregular polyhedron compounds. Other people, including Dirk Eisner, Leong Cheng Chit, Francesco Mancini, Hideaki Kawashima, Aaron

Pfitzenmier, and others, have contributed new designs over this period as well. The addition of hyperbolic distortion and the introduction of dihedral polar symmetry to Wire Frames have demonstrated that even more possibilities exist than previously thought. The complexity and technical difficulty of Wire Frames have increased dramatically since FIT, but the basic idea has remained the same.

When people contemplate actually folding these models, they typically say they are impossibly complex. I started out thinking the same thing, but with the right techniques, even the most difficult models become manageable. As you look through this section of the book, the directions may seem somewhat sparse, even vague, at points. This is because the goal of this book is not simply to have people follow instructions, but to create circumstances in which readers can work with only crease patterns for the units, and no instructions whatsoever for the assembly. This is quite possible; in fact, it is the key to understanding Wire Frames. Once the underlying symmetry of the model is determined, the weaving pattern can be intuited with a little practice. The pictures of the assembly are given more to help show the method of physically building each model than to actually show the weaving process exactly.

Additionally, while all of the units in this book have exact reference points

THINGS TO REMEMBER

» All paper proportions have been set up to work with an inch ruler. A metric ruler can be used as well, but the proportions should be enlarged as described in the Tips and Techniques section (page 7).

» The proportions can be modified very slightly, if necessary. Usually 1/16 (.0625) inch is an acceptable amount of modification, and shouldn't affect the model greatly.

» For best effect, sturdy paper should be used when folding Wire Frames with narrow paper proportions.

» If you intend to transport these models over long distances, it is best to fold them slightly tighter than recommended. For reasons not yet fully understood, the models tend to loosen during travel. I will leave it as an exercise to the reader to estimate how much the proportions should be reduced for such purposes.

» Paper proportions will always be rectangles, and can thus be approximated and ripped from squares with folded guidelines, or measured out with a ruler, marked with a pencil, and cut out. The latter is quicker, and allows you to customize the size of the final model as well, but often wastes more paper. If you decide to rip the paper from squares, however, keep in mind that the units must all be of the same width, so if they are proportioned, say, 1:2 and 1:5, the same-sized square can't be used for both unless horizontal divisions are added as well.

» The handedness of the units is interchangeable. In other words, the units can be folded in mirror image so that the pockets are on the side where the tabs currently are, and the tabs are on the side where the pockets currently are. As you become more comfortable with Wire Frames, you can switch the handedness to whichever orientation you prefer.

» You may also note that very few of the edge units in this volume have extra folds on the tabs—a common feature in many Wire Frame diagrams. This is because I have always found such folds to be superfluous, and they weaken the paper.

» As a general tip, it is preferable to assemble as many units as possible outside of the model. Units will always be easier to assemble outside of the model, and it is far easier the bend sections of the frames into the desired position than to assemble them in that position.

» Sometimes the angles may not seem to be mathematically perfect, but there are other factors to be accounted for. Some angles have to be wider or narrower than they would be in strict geometry—to make the locks stronger, for example; or to increase or decrease the dihedral angles, etc.

» Sometimes axial symbols will be shown for axes that aren't yet completed in the assembly pictures. These indicate where future axes will form on the model.

» All Wire Frame units start with the paper white side up, if the finished model is to be colored.

for the pockets and crimps, in many cases the folding process can be expedited by omitting the given reference steps and approximating the final folds. Again, the goal is to get aspiring folders to fully understand these models from a more autonomous perspective.

In addition to having similar units, another thing that sets Wire Frames apart from most other modulars is that they have a definitive number of units. Kusudamas and regular geometric modulars can usually be assembled in several fashions—tetrahedral, octahedral, and icosahedral, for example—and a different number of units would be required for each. This is not the case with Wire Frames, where the underlying symmetry of a model can only be made to align with a single polyhedron—for instance, a dodecahedron. A Wire Frame with dodecahedral symmetry cannot be assembled in an octahedral fashion. There is quite likely an octahedral version, but that would require different paper proportions and possibly different angles, and might not be woven together in the same fashion.

16 TRIANGLES

I CHOSE this model to be the first in the Wire Frame section because it is relatively easy, though it appears complicated. When I first designed it, I set out to make an eight-triangle compound by combining two different compounds of four triangles that each had tetrahedral symmetry. When I finished, I realized more triangles could be added, and thus increased the model from eight to sixteen triangles. This model has the fewest units of any Wire Frame in this book. It is also the only octahedral symmetry model in the book.

The paper proportions are 1:4. Ripping the paper from 6" squares makes a ~7" model.

⑴ Book-and-cupboard-fold and unfold.

⑵ Pivoting along the center crease, join the circled areas, folding where shown.

⑶ Swivel-squash-fold along the left cupboard fold line.

⑷ The result. Repeat on the top right side.

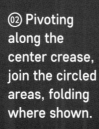

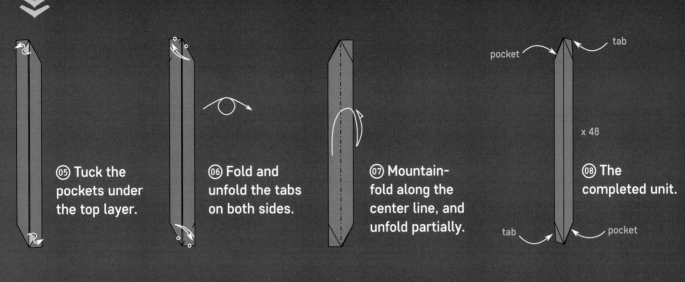

⑤ Tuck the pockets under the top layer.

⑥ Fold and unfold the tabs on both sides.

⑦ Mountain-fold along the center line, and unfold partially.

⑧ The completed unit.

pocket — tab

x 48

tab — pocket

ASSEMBLY

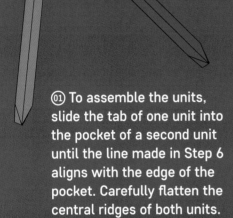

① To assemble the units, slide the tab of one unit into the pocket of a second unit until the line made in Step 6 aligns with the edge of the pocket. Carefully flatten the central ridges of both units.

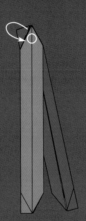

② Fold the opposite tab down and slide it into the other pocket. Be sure that the center crease of the top unit is flat where circled.

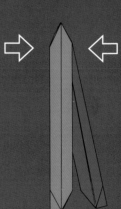

③ Making sure that both tabs stay in place, carefully apply pressure where shown, recreasing the central ridges.

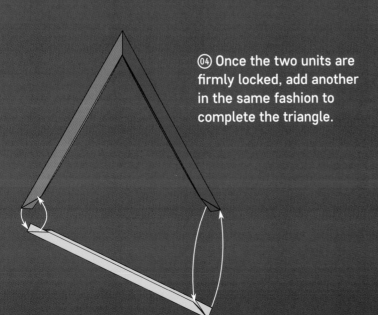

④ Once the two units are firmly locked, add another in the same fashion to complete the triangle.

⑤ One completed triangle. Make 15 more.

WEAVING INSTRUCTIONS

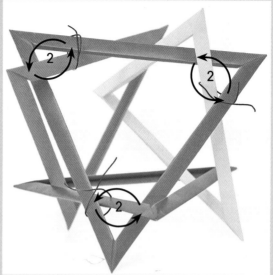

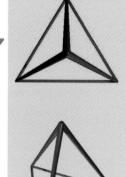

The first four frames will be the basic structure on which additional frames can be added. This set of four frames has tetrahedral symmetry. The upper right photo shows the two-fold symmetry, while the upper left shows the three-fold symmetry. Each triangle represents one face of a tetrahedron. It will be unstable at this point, so frame holders can be used as shown. Note where a three-fold axis appears in the picture on the left. There will be four of those as well, following the vertices of a tetrahedron. The two-fold axes, three of which are annotated on the right, will appear six times, aligning with the edges of a tetrahedron. By determining the symmetry of the shape, and its axes, you can begin to "intuit" the weaving in less visible areas, which will prove essential in many of the more complicated models presented later on in this volume.

Shown above are eight frames. Four have been added, changing the symmetry from tetrahedral to octahedral. Each triangle now represents one face of an octahedron. There are six four-fold axes, as shown above left, and eight three-fold axes, as shown above right. Note that each triangle has the same in-and-out weaving relationship with every other triangle except its opposite on the three-fold axis. An example of one such pair is the yellow and dark blue, as circled in the lower right picture on the opposite page. These two have the relationship shown in the below left diagram with every other triangle

except each other. As the symmetry of the shape becomes apparent, you should be able to guess where the three- and four-fold axes are on the other side of the model, using the octahedra below for guidance.

Adding the next set of four triangles, as shown in the images, requires a slightly greater understanding to weave correctly. At this point, the model has octahedral plus tetrahedral symmetry. Again, the set of four triangles is tetrahedral, and has four three-fold axes, which represent the vertices of a tetrahedron. New three-fold axes should appear inside of four of the eight existing three-fold axes. The circled frames are the new ones, the right picture showing one of the new three-fold axes, and the left showing the view

from the exact opposite three-fold. Although this can be a confusing step at first, the model should now be relatively sturdy if assembled correctly, and should be easier to work with.

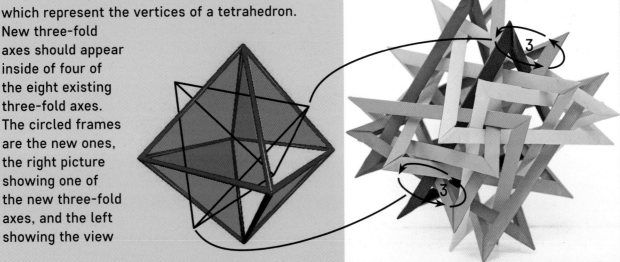

Assembling the last four triangles is probably the easiest part of the weaving process. The previous four triangles represented a tetrahedron poking through four faces of an octahedron; the remaining four represent another tetrahedron which will fill the remaining four three-fold axes, as shown on the bottom left. There are four sets of four parallel triangles, as circled on the right. The weaving process for every frame should follow an in-and-out weaving relationship—except for between any four parallel frames, as shown above. When in doubt about the weaving, you can use this rule to check. When finished, confirm that everything is correct. If you see bending or other distortion, there is probably an error somewhere. The schematic as shown below shouldn't really be necessary; it is included simply to illustrate the underlying geometrical connection.

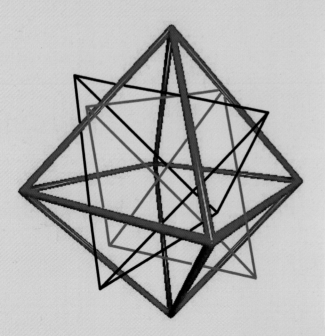

COSMOS

THIS MODEL is one of two in this book that belong to a near-infinite series of regular polygons wrapping around a sphere (the other is "The Alphabet," the last model in this book). This one roughly follows the symmetry of a truncated icosahedron, and as such, is one of the most spherical Wire Frames. Because of the wide interior angle of nonagons (140 degrees), standard units would be relatively weak, so upgraded ones are used here. They consume more paper, but have a larger tab and lock, and are stronger. This model is a nice, relatively simple introduction to icosahedral/dodecahedral symmetry.

This model requires paper proportioned 1:3.25. Starting with 3.5" squares makes a ~8.5" model.

⟫ MAKING THE TEMPLATE

① Fold in half, left to right.

② Fold the top layer out as shown.

③ Fold and unfold the top layer along angle bisectors.

④ The result. Unfold everything.

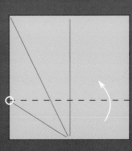

⑤ Fold up along the top of the crease made in the previous step.

⑥ The completed template.

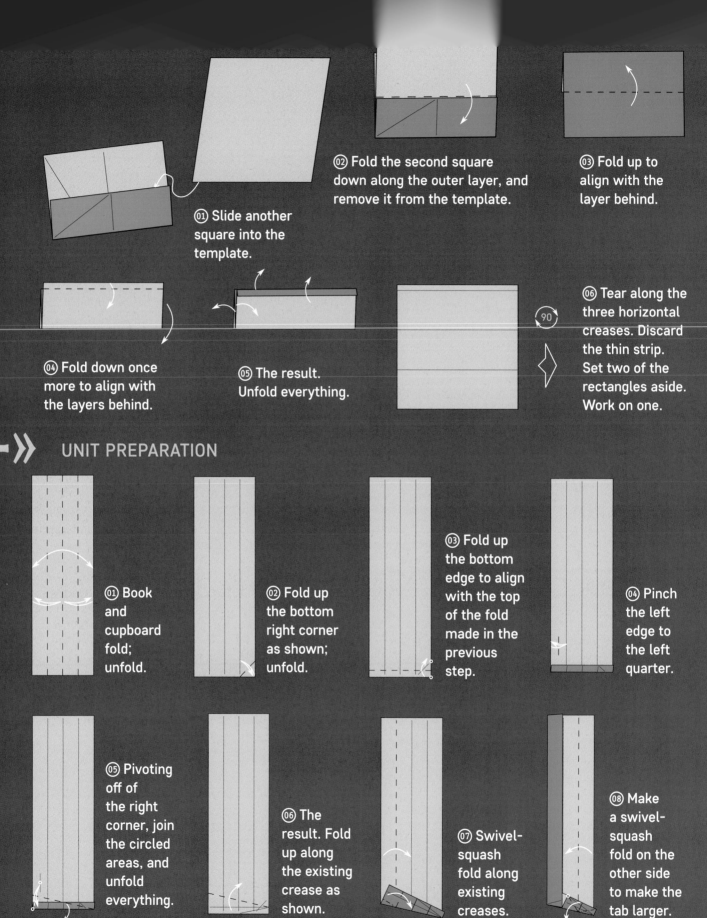

01 Slide another square into the template.

02 Fold the second square down along the outer layer, and remove it from the template.

03 Fold up to align with the layer behind.

04 Fold down once more to align with the layers behind.

05 The result. Unfold everything.

06 Tear along the three horizontal creases. Discard the thin strip. Set two of the rectangles aside. Work on one.

UNIT PREPARATION

01 Book and cupboard fold; unfold.

02 Fold up the bottom right corner as shown; unfold.

03 Fold up the bottom edge to align with the top of the fold made in the previous step.

04 Pinch the left edge to the left quarter.

05 Pivoting off of the right corner, join the circled areas, and unfold everything.

06 The result. Fold up along the existing crease as shown.

07 Swivel-squash fold along existing creases.

08 Make a swivel-squash fold on the other side to make the tab larger.

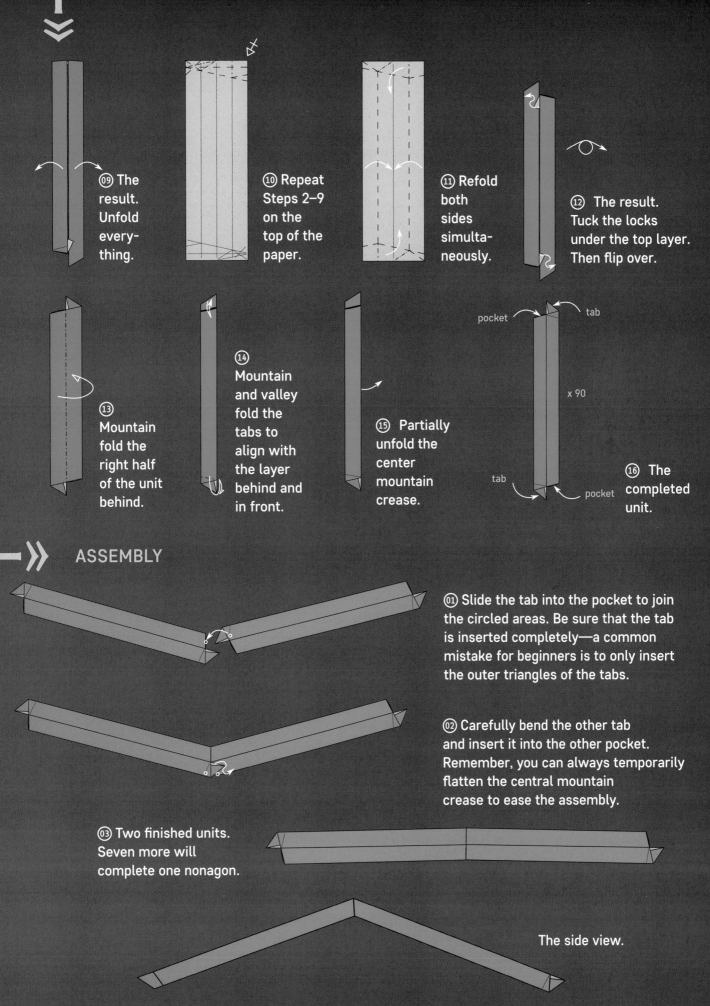

⑨ The result. Unfold everything.

⑩ Repeat Steps 2–9 on the top of the paper.

⑪ Refold both sides simultaneously.

⑫ The result. Tuck the locks under the top layer. Then flip over.

⑬ Mountain fold the right half of the unit behind.

⑭ Mountain and valley fold the tabs to align with the layer behind and in front.

⑮ Partially unfold the center mountain crease.

pocket   tab

x 90

tab   pocket

⑯ The completed unit.

ASSEMBLY

① Slide the tab into the pocket to join the circled areas. Be sure that the tab is inserted completely—a common mistake for beginners is to only insert the outer triangles of the tabs.

② Carefully bend the other tab and insert it into the other pocket. Remember, you can always temporarily flatten the central mountain crease to ease the assembly.

③ Two finished units. Seven more will complete one nonagon.

The side view.

WEAVING INSTRUCTIONS

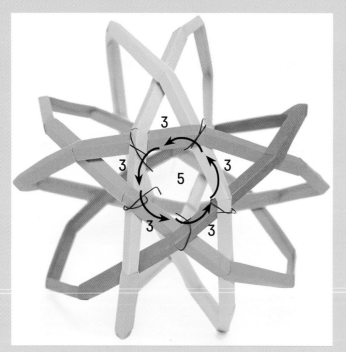

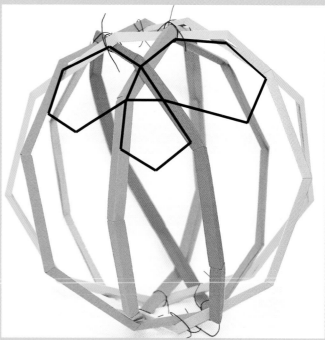

This model has an in-and-out weaving relationship between every frame, as shown in the diagram. Because the relationship between every frame is quite simple, the model is shown half completed in the pictures above. The five frames shown form two complete five-fold axes of the twelve that will be on the final model—one on either side as a polar opposite of the other. Around each five-fold axis will be five three-fold axes. Other than the five and three-fold axes on either side, there should be no interaction between the first five frames. You can add frame holders where shown. The picture on the upper right shows the side view. Around the three-fold axes will be five six-fold axes, which will be completed next.

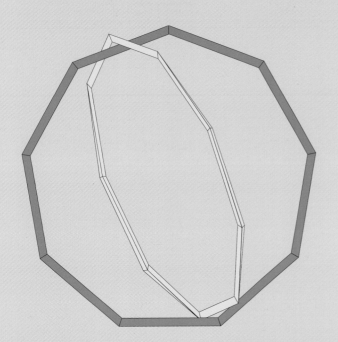

Be sure not to overtighten any frame holders, or the marks will show on the final model!

 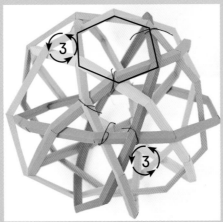 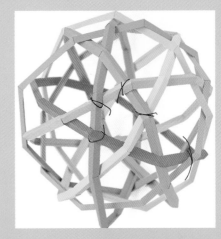

In the left-hand picture above, six frames are shown; in the center, seven; and on the right, eight. Adding the sixth frame to the model is probably the most difficult part of the assembly, mainly because the nonagons don't support themselves very well until the model is completed. Keep in mind that the model can be neatened up later, but if any units come apart completely, they should be fixed right away.

The last five frames all have the same relation-ship with the first five. The sixth frame forms a new three-fold axis on each side of the model, and what-ever relationship it has with the axis stays uniform until it reaches the other three-fold axis it formed. In other words, each new frame will be inside every other one of the previous five frames assembled, or outside of every other one, depending on which side of the model you look at. Upon the completion of seven frames, the first two six-fold axes will be completed, as shown in the central photo, above.

  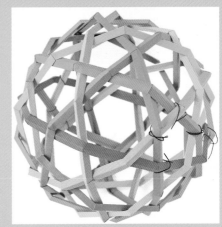

The last few frames are the easiest to weave (this is the case with most Wire Frames). Keep a trun-cated icosahedron in mind when assembling the last few frames, as the shape will really begin to become visible at this point. Of course, the shape isn't exactly a truncated icosahedron —there are no triangles in a truncated icosahedron as there are in this model. With twelve pentagonal five-fold axes and twenty hexagonal six-fold axes, however, there is definitely a similarity, and the triangles can easily be thought of as representing the vertices of the shape. The completed model is shown above in the center and on the right. Remove the frame holders, if any were used, upon completion.

NEBULA

THIS MODEL is the first in the book that is genuinely complicated to assemble. You will note that the geometric descriptions of the models are becoming longer and more unusual. I find the small pentagons that frame the five-fold axes to be the focal point of the model, although their original intention was simply to make the model stronger.

This model is also the first in the book for which a template is not really practical. While one could be made, it is easier to measure out and cut the required rectangles, which are proportioned 1.0625:7.5 and 1.0625:2.125.

THE LONGER UNITS, 1.0625:7.5, "A"

① Book and cupboard fold and unfold.

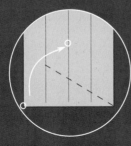

② Pivoting along the right edge, join the circled areas, folding where shown.

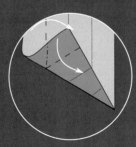

③ Swivel-squash fold along existing creases, folding the left quarter in.

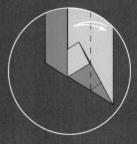

④ Fold and unfold the left quarter.

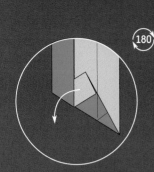

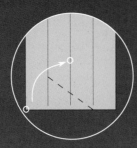

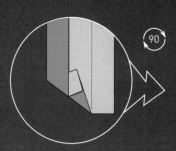

⑤ The result. Unfold everything, and rotate 180 degrees.

⑥ Pivoting along the right quarter, join the circled areas, folding where shown.

⑦ Swivel-squash fold along existing creases, folding the left quarter in.

⑧ The result.

⑨ Refold the right half of the unit along existing creases.

⑩ Fold the unit in half, bottom to top.

⑪ Fold the right tab down along the edge of the pocket; unfold.

⑫ Partially unfold the center crease.

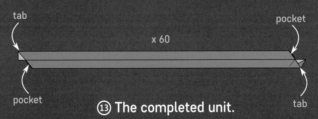

tab

pocket

x 60

pocket

tab

⑬ The completed unit.

THE SHORTER UNITS, 1.0625:2.125, "B"

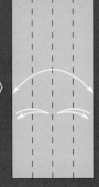

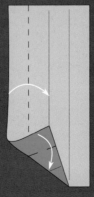

① Book and cupboard fold and unfold.

② Pivoting along the right quarter, join the circled areas, folding where shown.

③ Swivel-squash fold along existing creases, folding the left quarter in.

④ Unfold everything and rotate 180 degrees.

⑤ Repeat Steps 2–3.

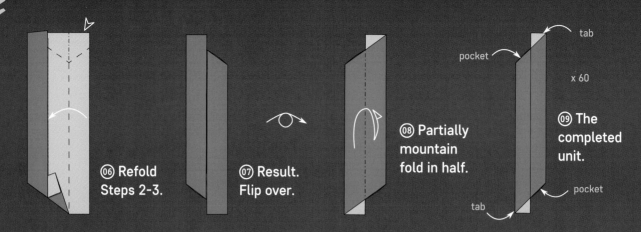

⑥ **Refold Steps 2-3.**

⑦ **Result. Flip over.**

⑧ **Partially mountain fold in half.**

⑨ **The completed unit.**

pocket

tab

x 60

tab

pocket

You may observe that the way these units are folded tends to be representative of Wire Frame folding in general. This is the qualifying characteristic of Wire Frame models. That most Wire Frames are woven compounds is not actually their determining feature.

ASSEMBLY

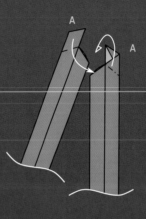

A

A

① **Any two longer units will be joined together with the 120-degree lock formed in Steps 2–4. Flatten the central ridges of the units down completely in order to assemble the units. Slide the tabs into the pockets at the same time. Then recrease the central ridges of the units to lock.**

② **To assemble the rest of the units, simply slide the tabs into the pockets. The shorter units have 108-degree angles on both sides, which is the measure of the interior angles of a pentagon. Two short units will be assembled into one longer unit. Make sure that the longer units' 108-degree angle is the one that is being assembled into the smaller units, which have 108-degree angles on both sides.**

B

B

B

A

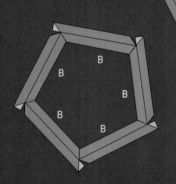

B

B

B

B

B

One assembled pentagon face, composed of five shorter units.

Always be sure to pinch the center ridges over the area where the tab is in the pocket.

③ **This is one completed frame—a polarly-truncated pentagonally distorted dipole. Each small pentagon forms the opposing poles, while the longer units join them together.**

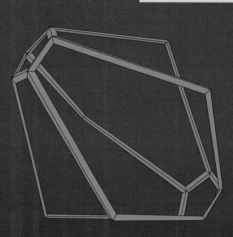

WEAVING INSTRUCTIONS

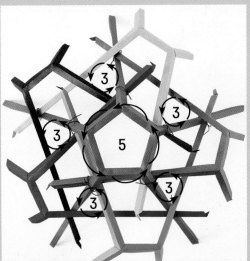

Despite this model's apparent complexity, if broken down into individual steps, it is easier than it appears. This model is an excellent candidate for the "bottom-up" weaving method, so it will be diagrammed here in that fashion. In the above pictures, thirty-five units have been assembled: ten orange units and five each of the remaining five colors. A five-fold axis composed of five different frames will form under each small pentagon in the model, such as the orange one in the pictures above. Around this will form five three-fold axes. The circles in the upper right photo show the limiting factors; these spots are where frame holders can be best used. Each long unit contributes one edge to a three-fold and a five-fold axis. As you might have expected by now, there will be twelve five-fold axes and twenty three-fold axes, aligning with the faces and vertices of a dodecahedron. The short units do not weave, and should always be on the exterior of the model. In the upper left photo, you can see the beginnings of the next five-fold axes. The light-green frame in the forefront is above parts of the red and orange frames, which will form the first two sides of the five-fold axis below.

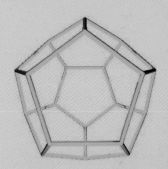

As the model progresses further, it should become physically easier to work with. Instead of looking at the weaving pattern of the whole model at once, which can be overwhelming, focus on a single small area of the model, and identify its relationship with everything else. When in doubt about the weaving, determine what that part of the model should look like. Once you are confident that a particular area of the model is woven correctly, you can reapply that pattern to other areas. In the photos to the right, sixty units are now assembled. The five half-completed pentagons from the first stage are complete; two additional long units have been added to the five frames from the first five-fold axis formed; and a single short unit has been added to those five frames as well, to begin the second pentagons for those five frames. Each of the surrounding five-fold axes (as shown to the right) should have four sides completed now. As you continue the model, you can flip it over, as shown on the following page, where eighty-five units have been assembled. By this point, the model should be relatively sturdy, so you can remove any frame holders that may be inhibiting your movement.

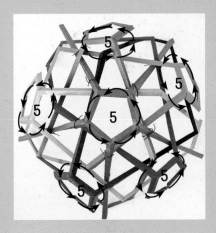

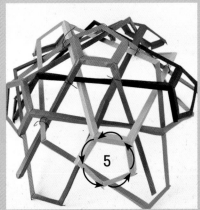

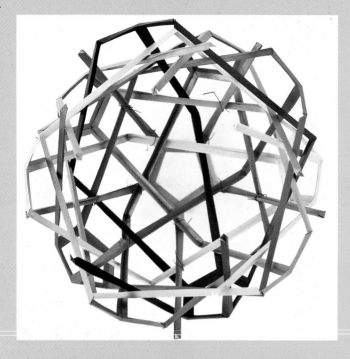

As the model nears completion, as in all other Wire Frames, the weaving pattern should be fairly self-evident. The lower six five-fold axes are now complete; the five above should be almost complete, with only one edge missing in each. Only the top five-fold axis remains—the one that will be under the orange frame, as in the beginning of the model. A common place for errors in all Wire Frames is in the weaving of the two-fold axes, one of which is circled in the above right picture, which roughly represent the areas between the three and five-fold axes. Note the white frame pokes out from underneath the adjacent five-fold axes, but is underneath the black, and blue-green frame, on either

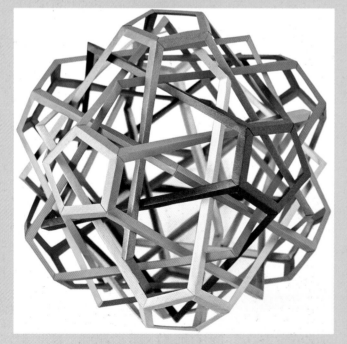

side of the point. It is an easy mistake to have these points, such as the white one, above the frames on either side, so be sure to check those carefully around the model. On the left is the completed model, as seen from the two-fold axis. Be sure to double check the weaving before pronouncing it complete!

ATMOSPHERE

THIS MODEL is similar to "Nebula," which precedes it. If you enlarged the small pentagons that formed the exterior and lowered them toward the center of the model, and added five more units to the equator of each frame, you would have this model. The name came from Daniel Kwan, who said the original design would have "more atmosphere" if thirty units were added.

The three different units are proportioned as follows: A) 1:7.875; B) 1:6.8125; and C) 1:2.6875. In inches, these proportions make a ~13"-diameter model. Start with the longest unit.

THE LONGEST UNITS, 1:7.875, "A"

① Book and cupboard fold and unfold.

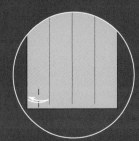

② Pinch the left edge into the left quarter; unfold.

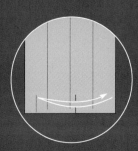

③ Pinch the right edge into the pinch just made; unfold.

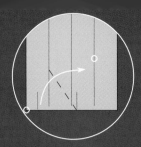

⑧ Refold the top of the unit.

④ Pivoting along the pinch made in the previous step, join the left corner to the right quarter.

⑤ Swivel-squash fold along existing creases, folding the left quarter in.

⑨ Join the circled edges on both sides, and then unfold.

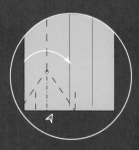

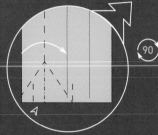

⑥ The result. Unfold everything and rotate 180 degrees.

⑦ Repeat Steps 2–4 on the other side of the paper.

⑩ Partially fold the unit in half, then flip over.

tab

pocket

x 30

pocket

tab

⑪ The completed unit.

THE MEDIUM-LENGTH UNITS, 1:6.8125, "B"

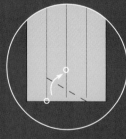

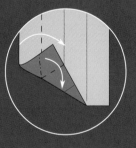

② Repeat Steps 2–5 from the longest unit. Rotate 180 degrees.

③ Pivoting along the right quarter, join the bottom of the left quarter to the center crease.

④ Swivel-squash-fold along existing creases, folding the left quarter in.

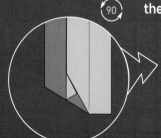

⑥ Refold the left of the unit.

① Book and cupboard fold and unfold.

⑤ The result. This is the 120-degree lock.

⑦ Join the circled edges on the left of the unit; unfold.

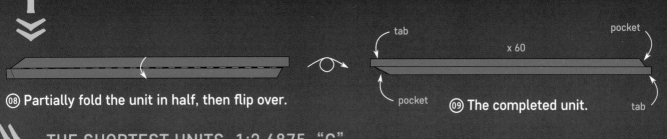

⑧ Partially fold the unit in half, then flip over.

tab

pocket

x 60

pocket

tab

⑨ The completed unit.

THE SHORTEST UNITS, 1:2.6875, "C"

① Book and cupboard fold; unfold.

180

② Repeat Steps 2–5 from the longest unit. Unfold everything, and then rotate 180 degrees.

③ Pivoting along the right quarter, join the left corner to the center crease.

④ Pivoting along the right quarter, join the left corner to the center crease. Swivel-squash fold along existing creases, folding the left quarter in.

⑤ The result. This is a 108-degree lock. Refold Step 2.

⑥ Partially fold the unit in half, then flip over.

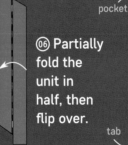

pocket

tab

x 60

tab

pocket

⑦ The completed unit.

ASSEMBLY

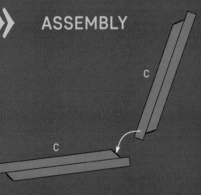

C

C

① This model is assembled much like the previous one. As before, simply slide the tabs into the pockets to lock. Five short units will join to form opposite pentagonal poles on each frame. Be very careful with the short units—the interior angles of a pentagon are 108 degrees, and it is critical that the 108-degree angles, as formed in Steps 3–5, are facing toward the interior of the pentagons. If any of the units start to bend or wrinkle when the last unit of the pentagon is added, one of them is probably facing the wrong direction.

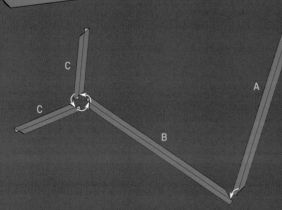

C

C

C

A

B

② The medium-length units will connect to the small pentagons through the 120-degree locks, and the longest units will attach to the other side of the medium-length units.

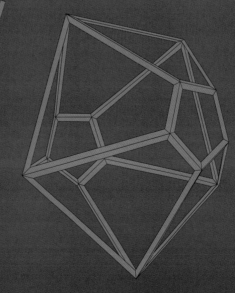

③ One complete pentagonal bifrustum. A frustum is any pyramid whose tip has been truncated along a line parallel to its base. In a bifrustum, opposite poles are truncated parallel to the shared equatorial polygon (in this case, a pentagon).

WEAVING INSTRUCTIONS

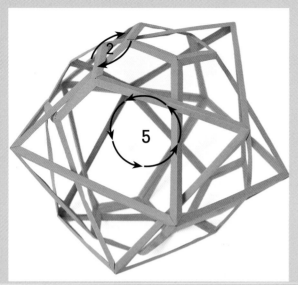

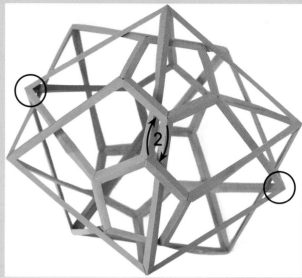

This model is assembled much like the previous one, "Nebula." Since the bottom-up weaving method was shown for that model, this one will be illustrated with the frame-at-a-time method. Beginning assembly can be overwhelming, so it is important to locate the axes as soon as possible. The two-fold view, as shown in the upper right picture, is the best position from which to start, because the two frames are symmetrical. Note the areas where the black circles are—on the left, the purple vertex is outside of the orange edge, while on the right, the orange vertex is outside of the purple edge. Although the opposite side of the model is difficult to discern, the weaving will be same, only in mirror image. The above left image shows the beginning of a five-fold axis. Each middle-length unit contributes on edge to one five-fold axis. As in the previous model, there will be a five-fold axis to correspond with both pentagonal faces of each

bifrustum. The below left image shows the beginning of a three-fold axis. Completed polyhedra don't have as great a degree of freedom to move around as polygons, so frame holders aren't as necessary here.

The third and fourth frames should proceed similarly to the first two. When adding new frames, each four-point vertex (each frame has five) will be on the outside of a two-fold axis. As with all models with dodecahedral symmetry, each two-fold axis will represent an edge of a dodecahedron. Below right and above left on the facing page, three frames are shown assembled; four are shown on the upper right and center of the facing page. Note how in each three-fold axis, three frames form a lower axis, while the three remaining frames form a second three-fold axis out of the longest units that compose each equatorial vertex. Both layers of each three-fold axis have the same chirality.

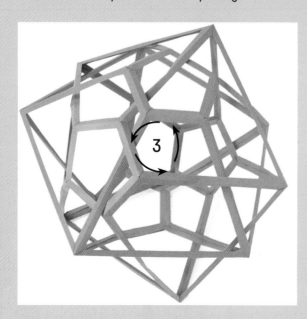

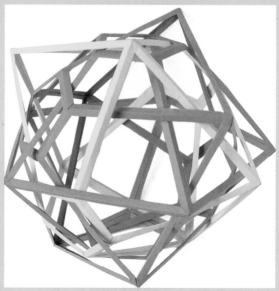

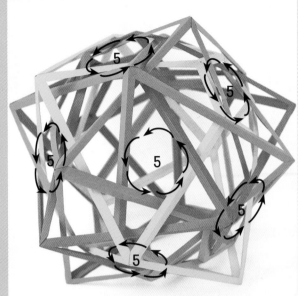

The internal vertices can be quite difficult to assemble, because they are not easily accessible. Take apart the outer vertices as needed to facilitate assembly. With five frames completed, the first two five-fold axes will be finished. In the bottom center picture, the two-fold view is shown. The relationship between any two frames in the model is best seen from the two-fold axis. The equatorial edges and vertices of each frame have a variation of the in-and-out weaving pattern. The bottom right photo shows the model with all six frames completed. The poles of each bifrustum frame the five-fold axes. While the symmetry of this polyhedron can be broadly defined as dodecahedral, it could also be related to an icosidodecahedron, or even a stellated rhombic triacontahedron. If the symmetry you see doesn't match the polyhedron given, feel free to find one that makes sense to you.

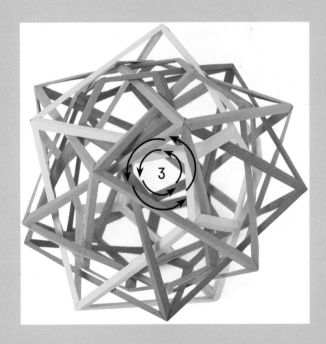

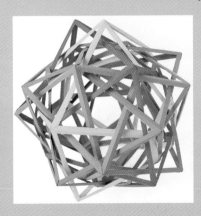

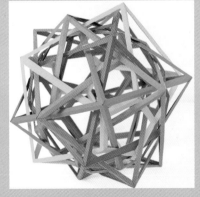

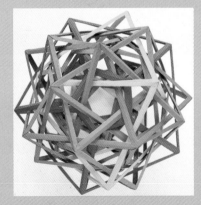

K5

THIS MODEL is a supremely attractive example of what the five-tetrahedron compound can become. It is similar to several other Wire Frame models, including Tom Hull's FIT and Robert Lang's K2. While each compound of four tetrahedra features the same weaving relationship as in Francisco Mancini's four-tetrahedron compound, their compilation is wholly original. The nickname comes from Dr. Lang's K2: I made another compound of twenty triangles and named it K3. The set expanded to include a number of other twenty-compounds of triangles/tetrahedra with dodecahedral symmetry.

This model requires paper proportioned 1:10. Translating the given dimensions into inches makes a model roughly 12" in diameter.

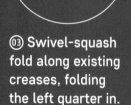

① Book-and-cupboard fold and unfold.

② Pivoting along the center crease, join the circled areas, folding where shown.

③ Swivel-squash fold along existing creases, folding the left quarter in.

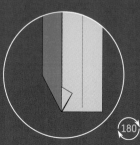

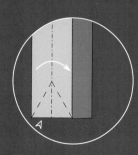

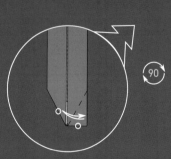

④ The result. Rotate 180 degrees.

⑤ Repeat Steps 2–4 on the other side of the paper.

⑥ Fold the tab in, joining the circled areas, and unfold. Repeat on the other side of the paper.

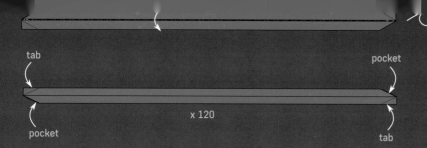

tab

pocket

x 120

pocket

tab

⑦ Partially fold in half, then flip over.

⑧ The completed unit.

>> ASSEMBLY

① Slide the tabs into the pockets and re-crease the central ridge of each unit to lock. Six units will assemble together—three at each vertex—to make one tetrahedron, as shown at right.

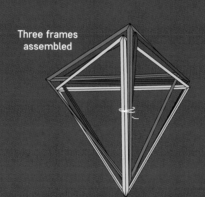

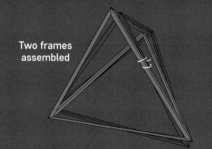

Two frames assembled

Three frames assembled

Four frames assembled

② Four individual tetrahedra will assemble together to make one "macro tetrahedron." Three of the four will form a three-fold axis, while the fourth will form a point over them. A different frame will have a point over each vertex of the tetrahedron.

③ You will probably want to use some kind of frame holder around at least one edge, as shown here.

④ This is one complete "macro tetrahedron." Be sure that the chirality of each three-fold axis is consistent.

WEAVING INSTRUCTIONS

Assembling this model requires the simultaneous coordination of two different weaving patterns: each four-tetrahedron compound, and five "macro-tetrahedra." Adding each new macro-tetrahedron requires each quad-beam edge to be thought of and assembled as a single edge. Wrap each quad-beam edge with a frame holder and assemble them as one unit. Weaving the five macro-tetrahedra in relation to each other is relatively simple, especially in comparison to other models in this book. Any

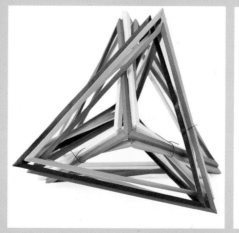

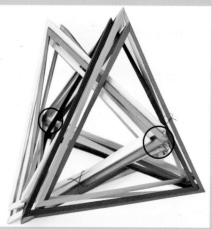

two macro-tetrahedra have the relationship shown above. Eight tetrahedra are assembled, and opposite vertices poke through opposite faces as circled.

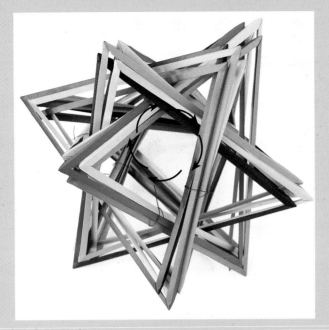

The two photos above show twelve tetrahedra (three macro-tetrahedra) from five- and three-fold views.

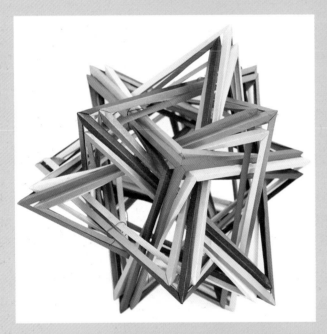
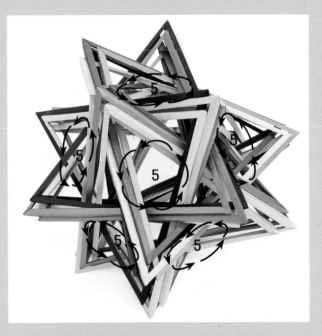

Assembling the remaining eight tetrahedra (the fourth and fifth macro-tetrahedra) is relatively straightforward. Envision a dodecahedron when assembling the five-fold axes. When adding frames, first weave the macro-tetrahedra together, and then carefully rework the frame so that each tetrahedron has the correct relationship with the others. If you have completed Tom Hull's FIT model, assembly of the macro-tetrahedra should be simple. Each edge contributes to two five-fold axes and two three-fold axes. In the above two photos, sixteen tetrahedra are shown assembled; at right, the completed model with all twenty tetrahedra is shown. The macro-tetrahedra shift without frame holders, so don't remove them until the model is finished.

GALAXY

THIS MODEL is best described in a word: "commanding." It is the second-largest model in this book by unit count, and it commands both great perseverance and, when completed, great attention. Each prism represents a face of a dodecahedron. It was designed during a study that several folders, including myself, were doing on polyhedron compounds with exterior faces that directly relate to the faces of a polyhedron. This particular model was the largest designed in that study.

This model has two different paper proportions: A) 1:3.375; and B) 1:4.8125. Translating the given dimensions into inches makes a ~10" model.

THE SHORTER UNITS: 1:3.375, "A"

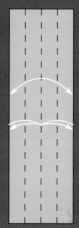

① Book and cupboard fold and unfold.

② Pivoting along the right quarter, join the circled areas, folding where shown.

③ Swivel-squash fold along the left quarter line.

④ The result. Unfold everything and rotate 180 degrees.

⑤ Pivoting along the right quarter, join the circled areas, folding where shown.

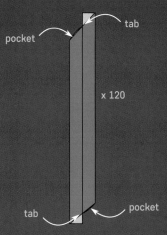

⑥ Swivel-squash fold along the left quarter line.

⑦ The result. Refold Steps 3–4.

⑧ Fold in half partially and flip over.

⑨ The completed unit.

pocket

tab

x 120

tab

pocket

THE LONGER UNITS: 1:4.8125, "B"

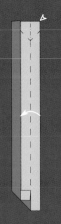
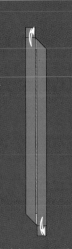
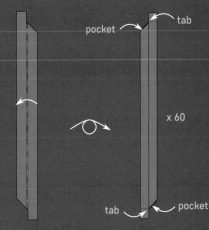

① Repeat Steps 1–4 as in the shorter units.

② Repeat Steps 2–3 as in the shorter units.

③ Refold the top of the unit.

④ Fold down the tabs along the top edge of the pocket, and unfold.

⑤ Fold in half partially and flip over.

⑥ The completed unit.

pocket

tab

x 60

tab

pocket

ASSEMBLY

The longer units have 90-degree angles on both sides, while the shorter units have one 90-degree angle and one 108-degree angle. The 108-degree angles are used to make pentagonal faces, while the 90-degree angles are used to make square/rectangular faces.

A

A

B

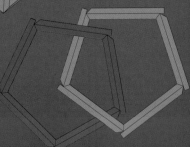

For reference, one complete frame, a pentagonal prism, should look like the image to the left.

Ease the difficulty of assembly by only assembling the shorter units into pentagons, setting the longer units aside for the moment. The model's base can be interpreted as six pairs of two pentagons each. Any two pentagons that are not a pair will have the in-and-out weaving relationship shown at right. Paired pentagons will not weave, but rather will be parallel. Thus, every pentagon will weave with ten others.

WEAVING INSTRUCTIONS

This model will be assembled as twelve "base" pentagons, followed by the addition of the longer edges and exterior pentagons. Two parts of the assembly process can be challenging: assembling the pentagons that form the base and assembling the longer units into the base pentagons—both of which are physical obstacles. At the top, five frames are shown assembled, with a sixth added in the two photos directly below. The unassembled side of the locks must be facing away from the center of the five-fold axes, of

which there will be twelve. Frame holders will be needed. One inner and one outer five-fold axis will form with the first five frames, and the sixth frame will "surround" the five-fold axes formed. Beginning the assembly will be physically difficult, so be prepared for an initial struggle.

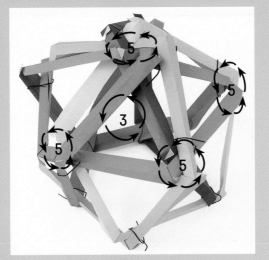

You need not be limited to assembling a Wire Frame in either the bottom-up or the frame-at-a-time styles. In the bottom two photos, I started with six complete frames, and then added three units each of five additional frames. I then proceed (see the following page) to add the next two sides for the five frames, and finally added the last frame, the parallel of the sixth that was added. Keep in mind that the base assembly will roughly represent an icosidodecahedron, and when combined

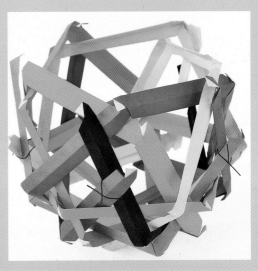

with the application of the in-and-out weaving pattern and the clearly delineated axes, the weaving shouldn't

be too difficult to figure out. In the photos directly above, three edges of the additional frames are added.

In the photo immediately to the right, the remaining two sides have been added to the frame from the previous page. The far right photo shows the complete base of twelve pentagons.

Once the base is completed, the remaining ten edges of each pentagonal prism can be added, either as one whole piece, or as individual units. Any two base pentagons that are pairs will weave into each other. For instance, in the photo immediately to the right, where one orange prism is completed, its pair, the black pentagon, will form the prism directly opposite the orange one, and its longer units will be over top of the base orange pentagon.

In the center right photo above, and the photo immediately to the right, six prisms are shown completed. The additions do actually weave; each longer unit will form one edge of one outer five-fold axis. Assembling the longer units into the base pentagons is probably the most difficult part of this model, because as you add more units, other frames will make it more difficult to manipulate them into place.

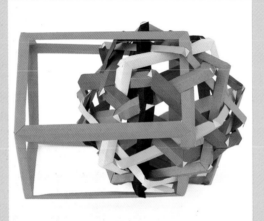

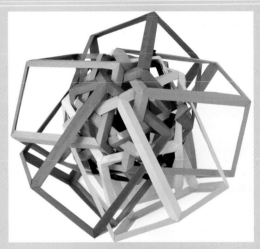

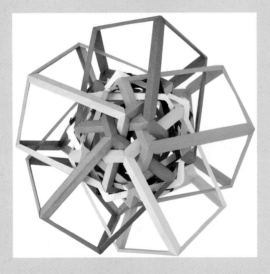

INTERSTELLAR

DESPITE ITS rather intimidating geometric name, this model is actually comparably easy to make. The combination of doubled five-fold axes with the small exterior triangles yields an attractive model which, because of its relatively wide paper proportions, can be folded at a smaller size. The exterior truncated points are only for aesthetics; they could be skipped, and the longer units extended to points, but the small triangles add great detail to the design, and are well worth the extra work required.

This model has two paper proportions: A) 1.375: 4.8125; and B) 1.375:1.5. Translating the given dimensions to inches makes a 6"-tall model.

THE LONGER UNITS, 1.375: 4.8125, "A"

① Book and cupboard fold and unfold.

② Pinch the left edge to the left quarter; unfold.

③ Pivoting along the right edge, join the circled areas, folding where shown.

④ Swivel-squash-fold along existing creases, folding in the left quarter.

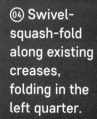

05 The result. Unfold everything, and rotate 180 degrees.

06 Pivoting along the right quarter, join the circled areas, folding where shown.

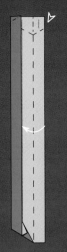

07 Swivel-squash fold along existing creases, folding in the left quarter.

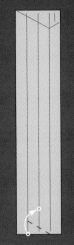

08 Refold Steps 3–4.

09 Crease the tab, joining the circled areas, and unfold.

THE SHORTER UNITS, 1.375:1.5, "B"

10 Partially fold in half, then flip over.

x 60

11 The completed unit.

pocket / tab / tab / pocket

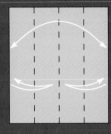

01 Book and cupboard fold and unfold.

02 Pivoting along the left quarter, join the circled areas, and unfold.

03 Join the circled areas, folding where shown.

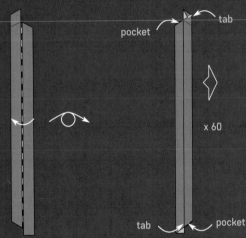

04 Swivel-squash fold along existing creases, folding in the left quarter.

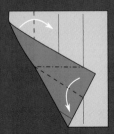

05 The result. Unfold everything and rotate 180 degrees.

06 Divide the third quarter from the left into thirds, pinching where shown. (See the note on the opposite page.)

 Steps 6 and 7 are not necessarily required; you can skip them if you wish and simply approximate the thirds. This will save time and remove unwanted creases from your model. However, the angles will be slightly less precise, unless you estimate perfectly. I rarely add precreases anymore, as I don't like them to be visible on my final model.

07 Divide the second quarter from the left into thirds, pinching where shown.

08 Pivoting along the left third pinch made in Step 6, join the circled areas, folding where shown.

09 Swivel-squash fold along existing creases, folding in the left quarter.

10 Refold Steps 3–4.

11 Fold in the tab along the base of the pocket.

12 Fold in half partially and flip over.

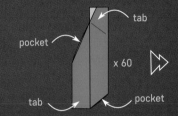

13 The completed unit.

ASSEMBLY

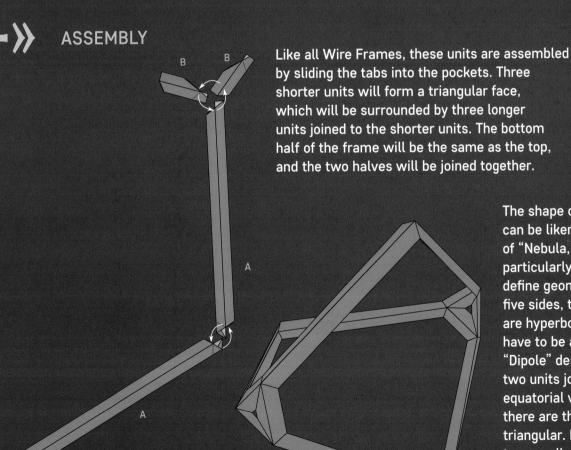

Like all Wire Frames, these units are assembled by sliding the tabs into the pockets. Three shorter units will form a triangular face, which will be surrounded by three longer units joined to the shorter units. The bottom half of the frame will be the same as the top, and the two halves will be joined together.

The shape of this frame can be likened to those of "Nebula," and thus is particularly difficult to define geometrically. It has five sides, three of which are hyperbolic, so we have to be a bit creative. "Dipole" describes the two units joining at each equatorial vertex, of which there are three, making it triangular. But there are two smaller triangles at each pole; thus, the frames are "polarly truncated."

WEAVING INSTRUCTIONS

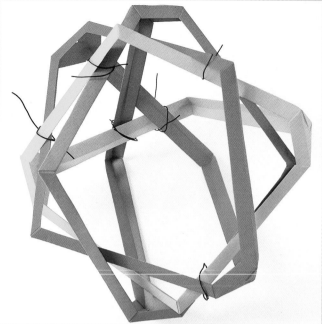

With its wider paper proportions and diminutive size, you might think that this model would be easy to assemble and weave. However, as there is no easily distinguishable relationship between any two frames, the weaving can be surprisingly complicated. In the upper photo at left, two frames are shown assembled; to its right, three. Below at left, four frames are assembled, while to its right, there are five. The assembly of the first five frames should form two different five-fold axes, each directly opposite the other. The axes shown on the left, above and below, are the smaller of the two; the axes on the right are the larger. Aside from axial weaving, there is little symmetry between the frames at this point. The use of frame holders is again nearly a necessity for this model.

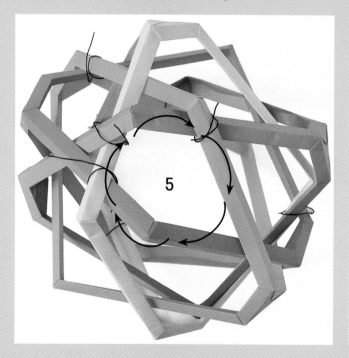

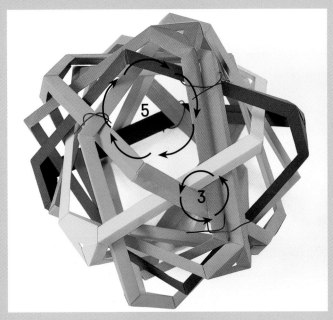

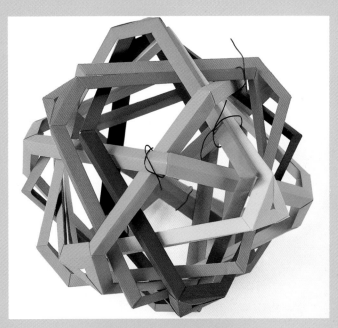

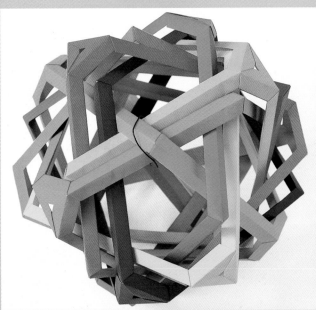

The top left photo shows six frames assembled. The top right photo shows seven frames assembled. Immediately above on the left there are eight frames assembled, and to the right of that, nine. While adding further frames to this model, keep in mind that the three-fold axes are doubled, so not every three-fold weave formed during assembly will be an axis on the final model. It is easiest to add each frame assembled as two identical halves, so that the equatorial vertices are the only ones that have to be connected on the model. Every long unit contributes one edge to a five-fold axis and one edge to an inner three-fold axis. The short units, of course, do not interact with any other frames. For the sake of space, I have not included a picture of the addition of the final frame, but I'm sure you'll know what to do!

DARK MATTER

THIS MODEL,

which is excellent for practicing the bottom-up weaving technique, combines crimped units with thin, large frames for an interesting airy effect. This is one of two complex, open models I designed that have twenty triangular faces, none of which interact with each other.

This model has two different paper proportions: A) 1:4.9375; and B) 1:6.625. Translating the given proportions into inches makes a model roughly 14" in diameter.

THE SHORTER UNITS, 1:4.9375, "A"

① A Book and cupboard fold and unfold.

② Pinch in half horizontally where shown.

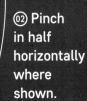

③ Fold the bottom up to the pinch; unfold.

④ Join the circled edges; unfold.

⑤ Fold an angle bisector through the circled intersection; unfold.

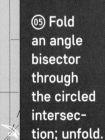

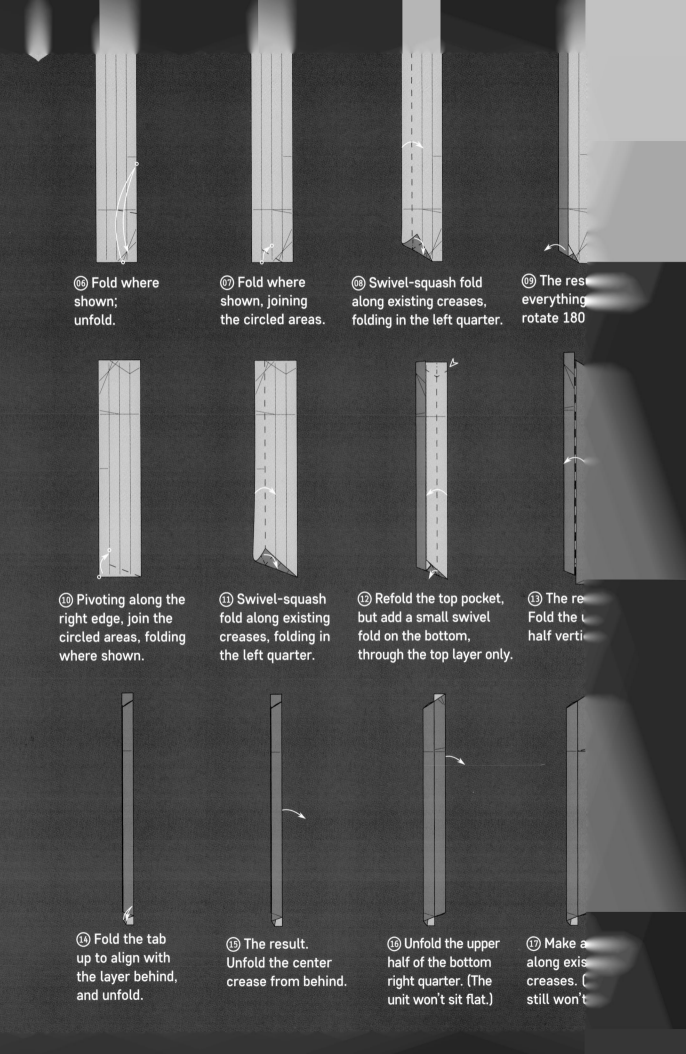

⑥ Fold where
shown;
unfold.

⑦ Fold where
shown, joining
the circled areas.

⑧ Swivel-squash fold
along existing creases,
folding in the left quarter.

⑨ The res
everything
rotate 180

⑩ Pivoting along the
right edge, join the
circled areas, folding
where shown.

⑪ Swivel-squash
fold along existing
creases, folding in
the left quarter.

⑫ Refold the top pocket,
but add a small swivel
fold on the bottom,
through the top layer only.

⑬ The re
Fold the u
half verti

⑭ Fold the tab
up to align with
the layer behind,
and unfold.

⑮ The result.
Unfold the center
crease from behind.

⑯ Unfold the upper
half of the bottom
right quarter. (The
unit won't sit flat.)

⑰ Make a
along exis
creases. (
still won't

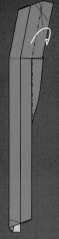

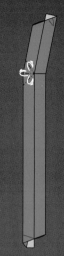

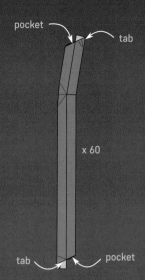

⑱ Mountain fold the quarter behind, allowing some paper in the crimp to stretch out and flatten.

⑲ Make three mountain folds where shown. (Their exact angle isn't important.)

⑳ Mountain fold the unit in half; partially unfold.

㉑ The completed unit.

THE LONGER UNITS, 1:6.625, "B"

① Book and cupboard fold and unfold.

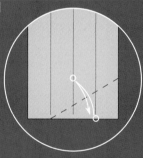

② Pivoting along the left quarter, join the circled areas, and unfold.

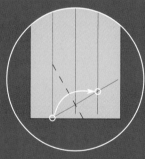

③ Join the circled areas, folding where shown.

④ Swivel-squash fold along existing creases, folding in the left quarter.

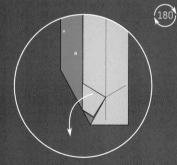

⑤ The result. Unfold and rotate 180 degrees.

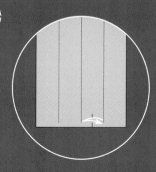

⑥ Pinch the right quarter to the center crease.

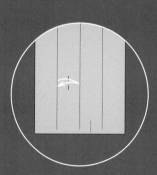

⑦ Pinch the left quarter to the center crease.

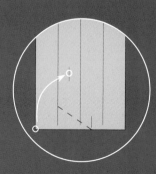

⑧ Pivoting along the pinch made in Step 6, join the circled areas, folding where shown.

⑨ Swivel-squash fold along existing creases, folding in the left quarter.

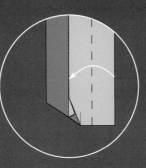

⑩ The result. Refold Steps 3–4.

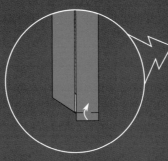

⑪ Fold the tab in along the base of the pocket.

⑫ Partially fold in half, then flip over.

pocket

tab

x 60

⑬ The completed unit.

tab

pocket

ASSEMBLY

Slide the tabs into the pockets, and recrease the center creases to lock. When locking the longer units into a shorter one, I recommend saving the triangular face locks for last, as they are the easiest to assemble.

B

A

A

B

One completed frame is shown at right. This shape is relatively complicated to explain geometrically; the name given comes from its folding origin, but it is easiest to visualize it as a wrinkled cuboctahedron with four tetrahedrically arranged faces replaced with three-point intersections.

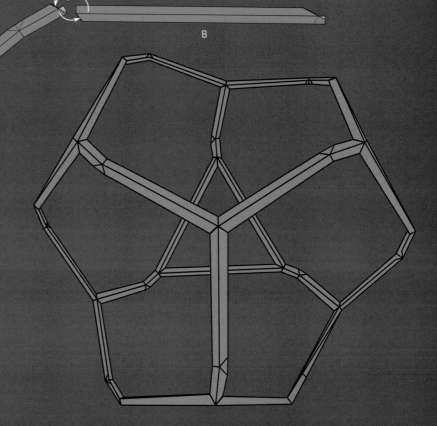

WEAVING INSTRUCTIONS

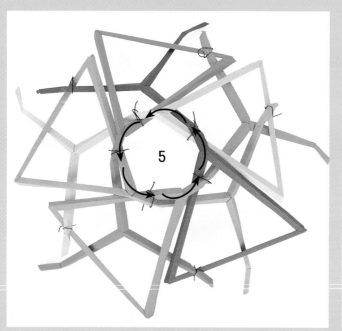

Because this model is a good candidate for bottom-up weaving, that process is shown in the pictures here. Once you become familiar with the general weaving technique, this model is actually much simpler than it initially appears. To begin, each of the triangular intersections will be directly under another triangular face. The triangular faces themselves will not interact with each other. Twenty three-point intersections join twenty free-floating triangular faces to complete the model. It is easiest to begin assembling the model from the outside, as shown in the above picture on the left, and once a few (around thirty or so) units are assembled, flip the entire model over, as shown above on the right, and assemble the remaining units. Below, sixty units are shown assembled in both photos.

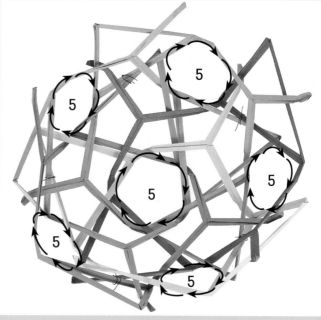

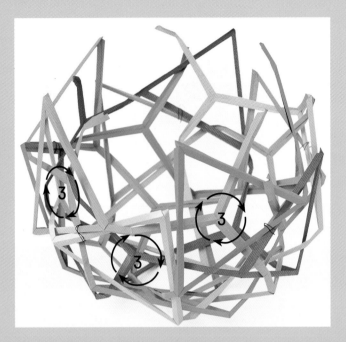

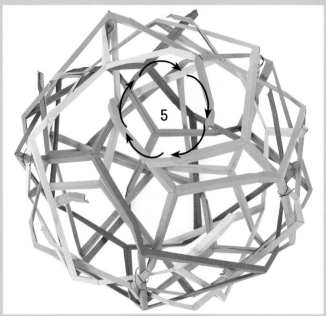

One of the great advantages of models whose vertices are all very near the perimeter (known as "surface-woven models") is that assembly is more convenient. Supporting the large open form of this model is a challenge that can arise upon assembly, however. It is helpful to have several medium-size objects handy to prop the model up while adding units. Above on the left, ninety units are shown assembled; above on the right, 105 units are shown assembled; and both pictures below show the model completed with all 120 units from two different views. As it nears completion, you can adjust the size of the five-fold axes slightly if necessary so that the crimps line up with the edges of the triangular faces with which they interlock.

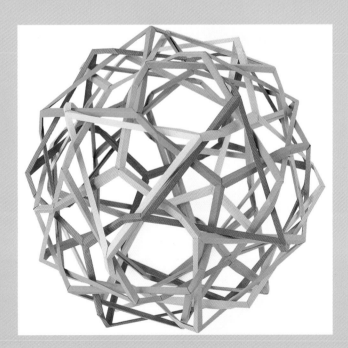

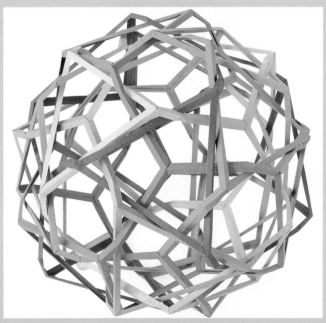

DARK ENERGY

THIS MODEL is the second of two models I designed with twenty non-interacting triangular faces. It is very similar in concept and folding to "Dark Matter," but the result is distinctly different. This has always been one of my favorites because the assembly is fun, and the paralleled edges around the five-fold axes give the model a complicated swirling effect. Although it does not appear similar, this model is a different weaving compound of the same type of shape as used in "Interstellar" (page 63).

There are two different paper proportions for this model: A) 1:6.5; and B) 1:7.75. Translating the given dimensions into inches makes a model ~14" in diameter.

THE SHORTER UNITS, 1:6.5, "A"

① Book and cupboard fold and unfold.

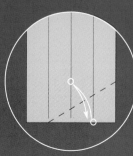

② Pivoting along the left quarter, join the circled areas; unfold.

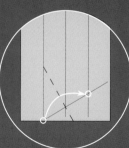

③ Join the circled areas, folding where shown.

④ Swivel-squash fold along existing creases, folding in the left quarter.

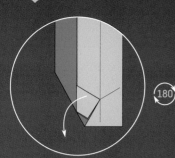

⑤ The result. Unfold and rotate 180 degrees.

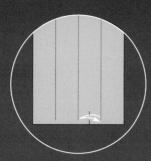

⑥ Pinch the right quarter to the center crease.

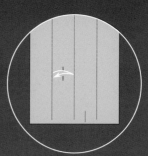

⑦ Pinch the left quarter to the center crease.

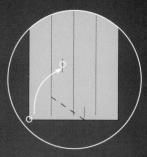

⑧ Pivoting along the pinch made in Step 6, join the circled areas, folding where shown.

⑨ Swivel-squash fold along existing creases, folding in the left quarter.

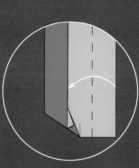

⑩ The result. Refold Steps 3–4.

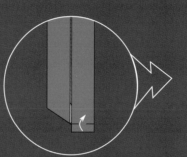

⑪ Fold the tab up along the base of the pocket.

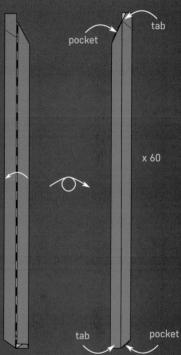

⑫ Partially fold in half, then flip over.

pocket

tab

x 60

tab

pocket

⑬ The completed unit.

THE LONGER UNITS, 1:7.75, "B"

① Book and cupboard fold and unfold.

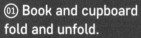

② Pivoting along the right quarter, join the circled areas, folding where shown.

③ Swivel-squash fold along existing creases, folding in the left quarter.

④ The result. Unfold and rotate 180 degrees.

⑤ Pinch the left edge to the left quarter.

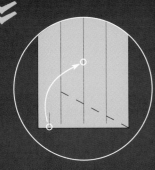

⑥ Pivoting along the right edge, join the circled areas, folding where shown.

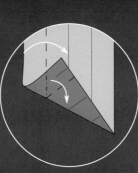

⑦ Swivel-squash fold along existing creases, folding in the left quarter.

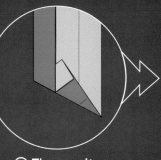

⑧ The result.

pocket   tab

x 60

tab   pocket

⑨ Refold Steps 2–3.

⑩ Fold the unit in half, right to left.

⑪ Mountain-fold the tab to align with the edge of the pocket behind.

⑫ Partially unfold the center crease from behind.

⑬ The completed unit.

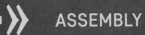

ASSEMBLY

As with all edge units, simply slide the tabs into the pockets to lock. The assembly here will be quite similar to that of "Dark Matter."

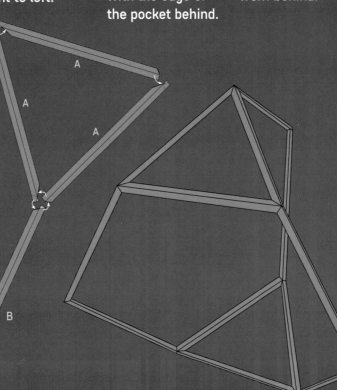

If you compare a completed frame from this model with a completed frame from "Interstellar" (a polarly truncated triangularly distorted dipole), you will see that they are the same shape. Calling this an equatorially diminished triangular bifrustum is just a matter of seeing the shape from a different geometric perspective.

WEAVING INSTRUCTIONS

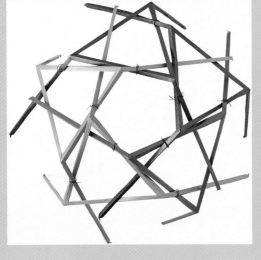

Any frame holders that are used during the assembly process should hold a shorter and longer unit together around a five-fold axis, rather than a two-fold axis, as the vertices over the two-fold axes are significantly higher than the surrounding frames. Keep the parallel longer edges in mind, as circled on the bottom left—if you switch them up, it can be confusing. In the middle row of photos to the left, seventy units are shown assembled; in the bottom row, eighty-five are shown. At this point, it is probably easiest to flip the model over so that the outside is facing down, as shown in the bottom right photo and on the following page. When in doubt as to which units have been added, track the path of one individual frame. Models like this can seem overwhelming until you become familiar with their weaving pattern.

As the model nears completion, you can remove frame holders and adjust any irregularities in the five-fold axes. The middle pair of images to the left show ninety units assembled, and the lack of interaction between any two triangular faces. All of the triangular faces should be assembled at this point. Whenever possible, try to assemble the two units that join at the equatorial vertices of every frame outside of the model. Assembling them as separate units that are already woven into the model can result in unsightly deformations in the paper. In the bottom two photos, the model is shown with all 120 units assembled from the three- and two-fold views. Before pronouncing the model complete, make sure that the pairs of parallel edges are both underneath the shorter edges of the two adjacent triangular faces above them.

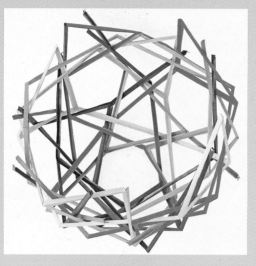

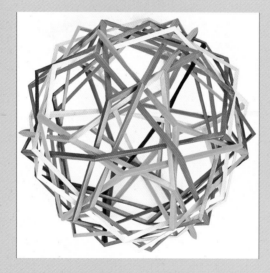

AURORA

THIS MODEL is surprisingly simple in concept, as the dipyramid is a more basic polyhedron than many others in this book. In fact I was surprised that no one had folded this before. Stellated compounds tend to be aesthetically pleasing; this one presents wonderful penta-grammoidal stars. The exterior vertices are difficult to assemble, and getting the units into the right areas can be tricky, though not nearly as much as in the model that follows this!

This model has two different paper proportions: A) 1.375:10.5, and B) 1.375:6.625. Translating the given dimensions into inches makes a ~12" model.

THE LONGER UNITS,
1.375:10.5, "A"

① Book and cupboard fold and unfold.

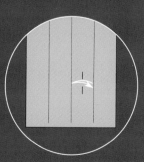

② Pinch the right quarter into the center crease where shown.

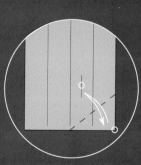

③ Pivoting along the center crease, join the circled areas; unfold.

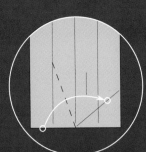

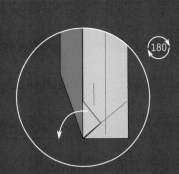

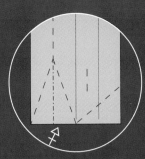

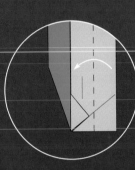

04 Pivoting along the center crease, join the circled areas, folding where shown.

05 Swivel-squash fold along existing creases, folding in the left quarter.

06 The result. Unfold everything and rotate 180 degrees.

07 Repeat Steps 2–5.

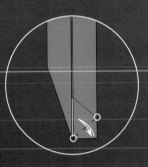

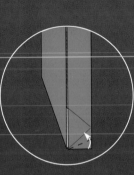

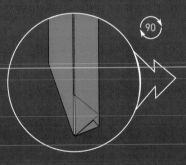

08 The result. Refold Steps 4–5.

09 Fold between the circled areas; unfold.

10 Fold in along the angle bisector of the previous fold.

11 Repeat Steps 9–10 on the top of the unit.

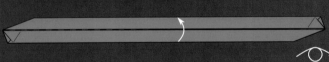

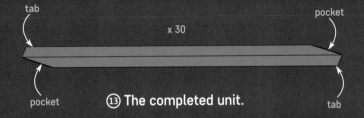

tab

pocket

x 30

12 Fold in half partially and flip over.

pocket

tab

13 The completed unit.

THE SHORTER UNITS, 1.375:6.625, "B"

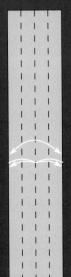

01 Book and cupboard-fold and unfold.

02 Repeat Steps 2–6 as in Unit A.

03 Pinch the left edge into the left quarter, folding where shown.

④ Pivoting along the right quarter, join the circled areas, folding where shown.

⑤ Swivel-squash fold along existing creases, folding in the left quarter.

⑥ Refold Step 2; do not unfold.

⑦ Tuck the pocket underneath the top layer.

⑧ Repeat Steps 9–10 on the top as in Unit A.

Steps 9-11 aren't actually required for Unit A in order for the units to be assembled, but they make the assembly much easier by reducing the size of the tabs. If you are looking for an extra challenge, you can omit these steps.

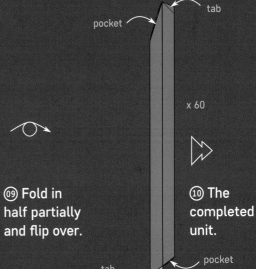

pocket

tab

x 60

tab

pocket

⑨ Fold in half partially and flip over.

⑩ The completed unit.

ASSEMBLY

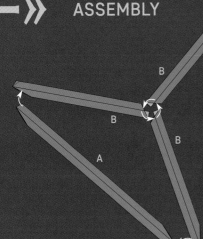

B

B

B

A

To assemble the units, slide the tabs into the pockets. Three shorter units will join together to make each of the two polar vertices; two shorter and two longer units will join alternately to make each of the three equatorial vertices. The equatorial vertices can be particularly challenging to assemble because of their narrow 20-degree angles.

Each frame in this model is a triangular dipyramid; that is, a six-sided polyhedron composed of triangles. In its regular form, a triangular dipyramid is a member of the Johnson solids polyhedron set.

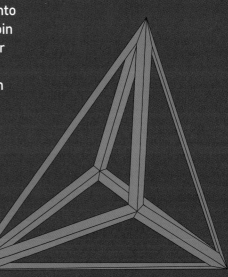

WEAVING INSTRUCTIONS

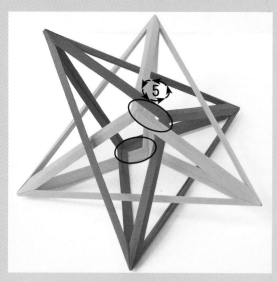

The assembly of this model can be quite challenging, especially physically, so it is important to follow several general points. First, each of the two polar vertices on each triangular dipyramid represent the center of one three-fold axis, and they should always be assembled beforehand. There are two triangular vertices per frame, and ten frames. This results in twenty three-fold axes, which, of course, align with the faces of an icosahedron. The polar vertices are close to the center of the model, so this definitely isn't a candidate for bottom-up weaving.

To the left, the photo shows two frames; the below left photo shows three. The photo directly below shows four, and the bottom one shows five. Note that between any two-fold axis "pairs," there is a sort of hybrid in-and-out weaving pattern.

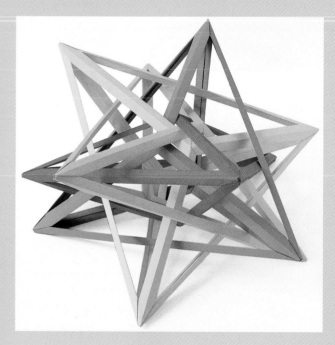

It is imperative to "intuit" the weaving process in this model, as the dense center obscures the vision. The trickiest part of the assembly is, undoubtedly, the addition of the triangular vertices near the center of the model. They can't be assembled inside the model, so they must be assembled outside of it and then carefully flattened and bent so that the three outer edges of the units nearly touch. The pieces form a spearhead shape when bent, which will allow you to push the point in; you can then unbend the units inside the model and adjust them accordingly. This process seems complicated, but it is the easiest way I have found to carry out the assembly.

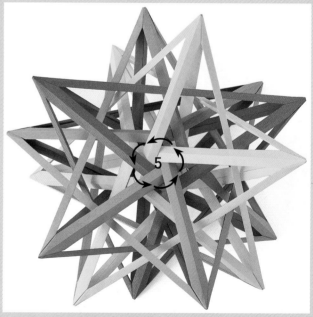

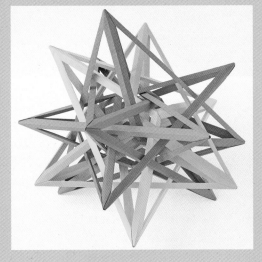

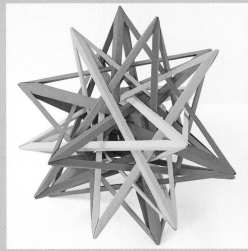

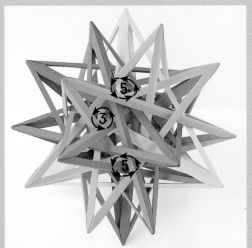

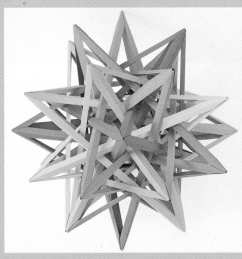

Directly to the left, six frames are shown; to the right of that, seven; in the middle left photo, eight frames are shown; and to the right of that, nine are shown. The bottom left photo shows the model from the three-fold view with all ten frames completed; the bottom right photo shows a two-fold view of the same.

As the model nears completion, check that all adjacent inner vertices have the two-fold hybrid in-and-out weaving relationship mentioned previously, and be sure the outer equatorial edges of each dipyramid contribute one edge to two different exterior five-fold axes. Confirm that the exterior stellated vertices are all assembled as cleanly as possible, since they are highly visible. After checking everything, the model can be pronounced complete! This is a nice prequel to the next project, which is similar, but more difficult.

EVENT HORIZON

THIS MODEL IS the most difficult one to assemble in this book. It is so dense that almost no light passes through, and is even more stellated and complex than "Aurora." Despite this, it is one of my favorite compounds. You should have a thorough understanding of Wire Frames before attempting to fold it.

This model has two different paper proportions: A) 1.125:6.1875; and B) 1.125:9.5625. Translating the given dimensions into inches results in a ~10" model.

THE SHORTER UNITS, 1.125:6.1875, "A"

① A Book and cupboard fold and unfold.

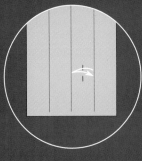

② Pinch in half between the center crease and the right quarter.

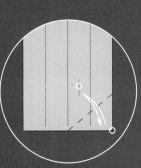

③ Pivoting along the center crease, join the circled areas; unfold.

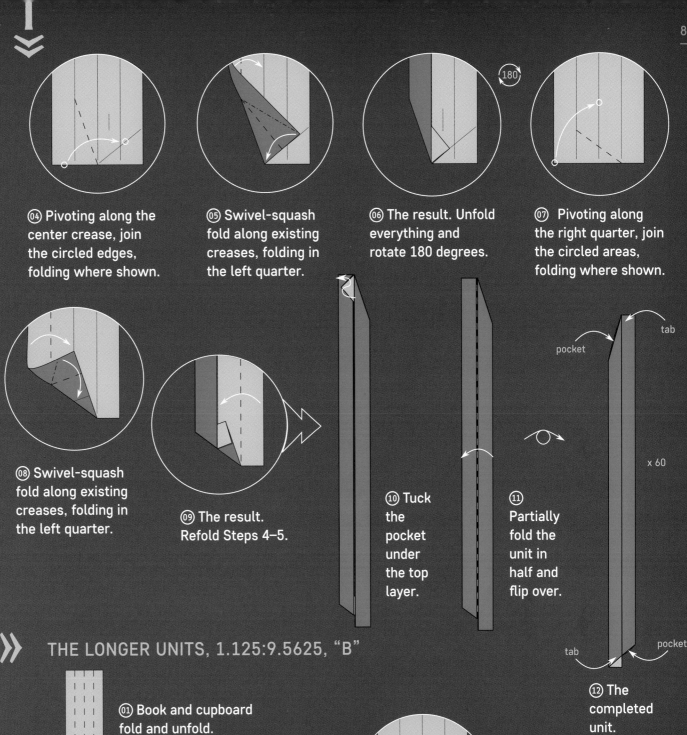

④ Pivoting along the center crease, join the circled edges, folding where shown.

⑤ Swivel-squash fold along existing creases, folding in the left quarter.

⑥ The result. Unfold everything and rotate 180 degrees.

⑦ Pivoting along the right quarter, join the circled areas, folding where shown.

⑧ Swivel-squash fold along existing creases, folding in the left quarter.

⑨ The result. Refold Steps 4–5.

⑩ Tuck the pocket under the top layer.

⑪ Partially fold the unit in half and flip over.

tab

pocket

x 60

tab

pocket

⑫ The completed unit.

THE LONGER UNITS, 1.125:9.5625, "B"

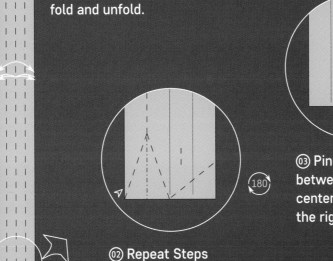

① Book and cupboard fold and unfold.

② Repeat Steps 2–6 as for Unit A.

③ Pinch in half between the center crease and the right quarter.

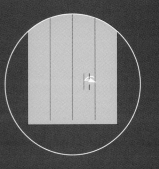

④ Pinch in half between the pinch just made and the right quarter.

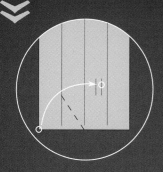  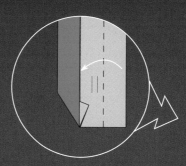

⑤ Pivoting along the center crease, join the circled edges, folding where shown.

⑥ Swivel-squash fold along existing creases, folding in the left quarter.

⑦ Refold Step 2. Don't unfold or rotate the paper.

pocket   tab

⑧ Tuck the pockets under the top layers.

x 60

⑨ Pivoting along the center crease, join the circled edges, and unfold.

⑩ Fold in half partially and flip over.

⑪ The completed unit.

tab

pocket

ASSEMBLY

To assemble the units, slide the tabs into the pockets, and recrease the center mountain folds to lock. The three lower vertices are most easily assembled in the numeric order shown below.

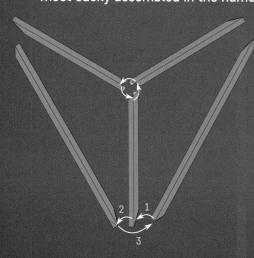

One complete frame is shown at right. However, it is easier to assemble the frames into the model in two separate steps, so complete frames should not be assembled outside of the model. Instead, assemble the three longer units of each frame into triangles, and the three shorter units of each frame into three-point intersections.

WEAVING INSTRUCTIONS

This model is best suited to a weaving method called "scaffolding." Frame-at-a-time weaving would be extremely complicated, and the interior vertices reach too far toward the center of the model to make bottom-up weaving practical. Basically, the model is built in separate sections—first the twenty triangles that form the bases of each tetrahedron, then the triangular points used to complete the tetrahedra. In the top pair of photos, five triangles are shown assembled. This creates two five-fold axes: one exterior one on the top, and a lower one on the bottom.

The middle pair of photos show ten triangles. The next five triangles form the same shape as the first five, but are flipped to opposite poles. Make sure the open sides of the triangles face outwards.

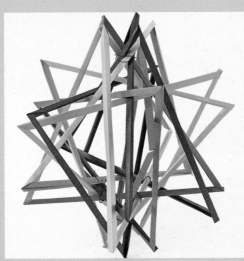

The bottom two photos show fifteen triangles assembled; on the following page, all twenty are shown. Frame holders will be needed, and are best applied as a single piece to group each of the twelve exterior five-fold axes. Each unit contributes one edge to one exterior five-fold axis, one interior five-fold axis, and one three-fold axis. The frames should be assembled with as much precision as possible. It doesn't matter if there are a few weaving mistakes; they can be fixed upon the addition of all twenty triangles. Frames 11 through 15 will form a third, wider five-fold axis around the original top five-fold axis, as shown on the bottom right.

These frames also contribute the interior edges of the top three-fold axes. Upon adding all twenty triangles, carefully check that everything matches the pictures in the top row to the right.

The relationship between any two inner three-point segments is shown in the left-hand photo in the middle row. Each point will be at the center of a three-fold axis, and any set of two will contribute two five-fold axes where shown. In the right-hand photo in the middle row, the model is shown with five inner points assembled. Assembling the inner points is extremely difficult, especially the last few. The bottom left photo shows fifteen three-point segments assembled, with a sixteenth being added. As in "Aurora," each point should be carefully bent and slowly adjusted into place. This is very time-consuming; figure on spending at least ten minutes adding and assembling each piece properly. The bottom right photo shows the completed model from the two-fold view.

THE ALPHABET

THIS IS the single most time-consuming project in this book; when it was first constructed it was the largest Wire Frame modular ever made. Its polar dihedral symmetry distinguishes it from all other models in this book, as it is not based on a regular polyhedron. The name comes from its connection to planar modulars, where each plane is a assigned a letter of the alphabet. Here, the twenty-six frames represent each of the letters of the alphabet.

This model has only one paper proportion: 1:2.41. If using the template, start with relatively small squares. Rectangles ripped from 3.5" squares make a ~10" model.

» MAKING AND USING THE TEMPLATE

① Book fold and unfold.

② Fold in half diagonally where shown; unfold.

③ Fold in half between the circled areas where shown; unfold.

④ Fold the right edge to join the intersection of the two previous folds; unfold.

⑤ Fold the right edge to join the previous fold; unfold.

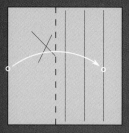

⑥ Fold the left edge to join the previous fold.

⑦ The completed template.

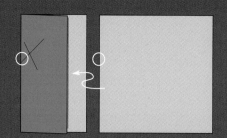

⑧ Slide a second square into the template, so that the circled lines are flush with each other.

⑨ Fold the square over along the edge of the template.

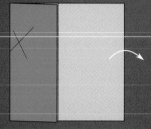

⑩ Pull the square sheet out and set the template aside.

⑪ Refold the crease.

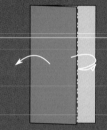

⑫ Mountain fold the edge to align with the top layer, and unfold everything.

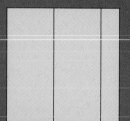

⑬ Rip along both lines. Keep the two equal-sized rectangles; discard the small strip on the right. Set one rectangle aside and focus on the other.

UNIT PREPARATION

The crimp formed using the given folding method makes a 75-degree interior angle. Such units are used to make stellated polygons. The actual angle for perfectly flat polygonal frames as used here should be roughly 74.4 degrees. Using the formula ($360/n+a$), where n is the number of sides on the given polygon and a is the interior angle of each vertex, you can find the necessary crimp angle for any star polygon. The 75-degree angle lets the units have a slightly wider dihedral angle than perfectly angled crimps would allow.

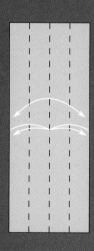

① Book and cupboard fold; unfold.

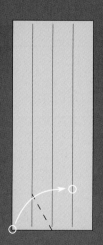

② Pivoting along the center crease, join the circled areas, folding where shown.

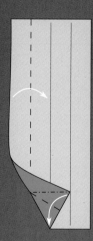

③ Swivel-squash fold along existing creases, folding the left quarter in.

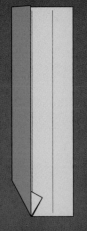

④ The result. Rotate 180 degrees.

⑤ Repeat Steps 2–3 on the other side of the paper.

⑥ Tuck the locks under the top layer.

⑦ Join the circled edges, and unfold.

⑧ Fold the unit in half horizontally; unfold.

⑨ Join the circled edges and unfold.

⑩ Fold up where shown, along the bottom of the crease made in the previous step; unfold.

⑪ Fold an angle bisector of the previous two steps.

⑫ Pivoting along the horizontal center crease, join the left edge to the edge of the angle bisector on the right, folding where shown; unfold.

⑬ Fold between the circled areas; unfold.

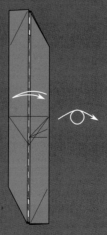

⑭ Fold in half and unfold. Then flip over.

⑮ Crimp the unit as shown, mountain folding the unit in half vertically.

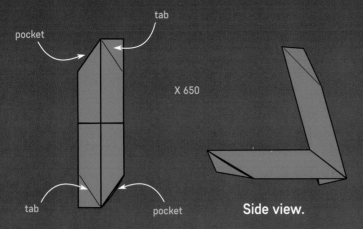

pocket
tab
tab
pocket

X 650

⑯ The completed unit.

Side view.

Steps 9–11 are used to make a 75-degree crimp angle. If you feel comfortable doing so, you can omit these steps and approximate the fold made in Step 12 to save time.

ASSEMBLY

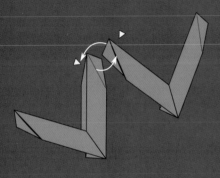

① Before assembling the units, flatten the dihedral angle around the vertices of the units you are assembling. Then slide the tabs around and into the pockets. Two units will join at every vertex.

② Finally, push in the sides of the units where shown, recreasing the center dihedral angle.

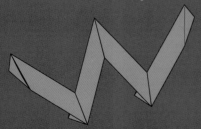

③ Two units joined. Join the others in the same fashion.

④ One complete icosikaipentagonal star, composed of twenty-five units. The name sounds intimidating, but it is simply the geometric name for a twenty-five sided polygon. It is called a star because it is a stellated polygon rather than a regular one, but is not an icosikaipentagram. Thus this model is composed of twenty-six interlocking twenty-five-sided stars.

WEAVING INSTRUCTIONS

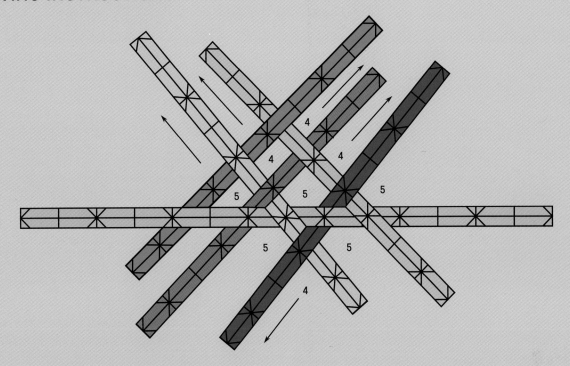

This model is very different from all others in this book. First, it is critical to recognize that it has polar dihedral symmetry; there will be a large "hole" on the top and bottom of the model. There is no regular polyhedron on which this compound is based; therefore the assembly will rely on axial weaving. Basically, there are twenty-five polygonal "bands" that weave around each other to form the large "holes" on either side of the model. The twenty-sixth frame wraps around the center of the model, and is

the only frame that has no interaction with the poles. In the photos below, the twenty-sixth frame is the light-blue band that is fully completed, and the pieces of other frames are parts of the center. The weaving of this model is actually simpler than you might think, once the pattern is understood. The schematic above shows the center light-blue frame—represented by the light-blue segment—and its equatorial relationship with other frames. There will be ten four-fold axes along each "file" of the model.

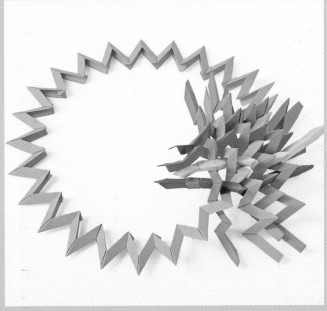

The actual weaving between the frames should be obvious—each vertex where two units join should weave over the "valley" crimp of another unit. Along each "file" of the model, starting at the center twenty-sixth frame, there will be one five-fold axis, followed by ten four-fold axes, followed by one three-fold axis near one of the poles, before the frame bends back down into the model as indicated by the arrow on the photo at the top left of the facing page.

This pattern is repeated along each "file" around the entire model. All frames, with the exception of the twenty-sixth center frame, are hyperboloidal. The number of assembled units in each picture is not specified since units can be added randomly, as needed. Also, the nature of the crimped units should make frameholders unnecessary, even in the earlier stages. The below right and top left facing page photos show the first complete pole forming.

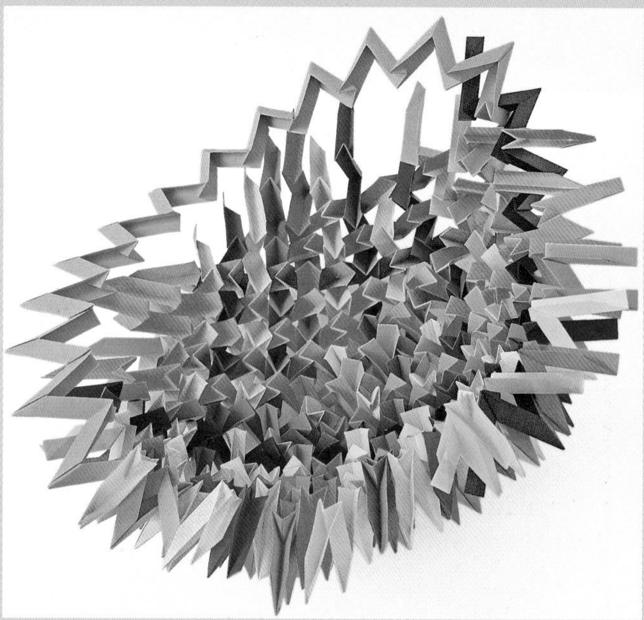

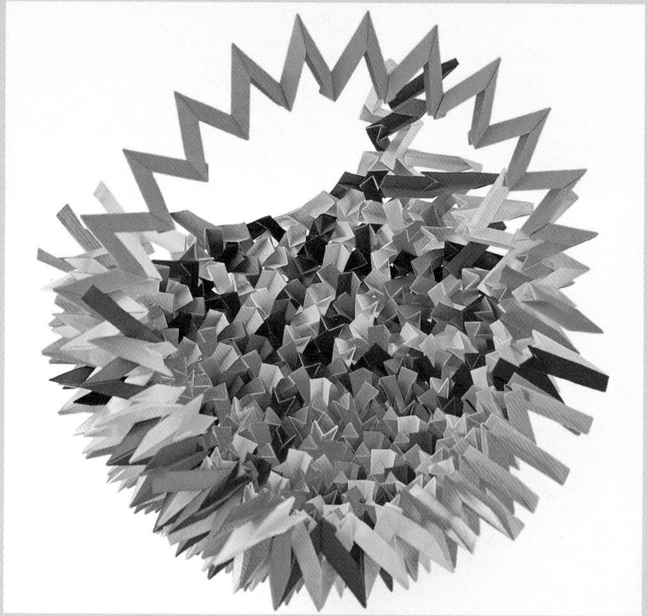

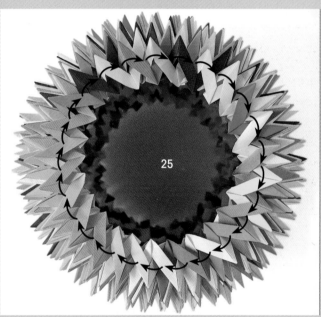

As the model gets closer to completion, its weight and shape will necessitate that it be tipped onto one of its poles (the large open holes with twenty-five-fold axes on opposite sides of the model). As the frames approach completion around the equatorial regions, the light-blue twenty-sixth frame might deform a little because the other frames still aren't in their final positions. Push down gently on the poles to help alleviate this problem. Additionally, note that along each "file," the chirality of every axis is opposite to the ones on either side of it. Further, this model has an in-and-out weaving relationship between any two frames, which is helpful if you need to double-check the weaving. If you manage to complete it, congratulate yourself; this is the largest and most time-consuming model in this book, which is why I have left it for last.

Published by Tuttle Publishing, an imprint of Periplus Editions (HK) Ltd.

www.tuttlepublishing.com

Library of Congress Cataloging-in-Publication Data

Loper, Byriah, 1994- author.
 Mind-blowing modular origami : the art of polyhedral paper folding / Byriah Loper.
    pages cm
 ISBN 978-4-8053-1309-1 (paperback)
1. Origami. I. Title.
 TT872.5.L67 2015
 736'.982--dc23
                         2015017494

ISBN 978-4-8053-1309-1

DISTRIBUTED BY

NORTH AMERICA, LATIN AMERICA & EUROPE
Tuttle Publishing, 364 Innovation Drive, North Clarendon, VT 05759-9436 U.S.A.
Tel: (802) 773-8930 | Fax: (802) 773-6993
 info@tuttlepublishing.com | www.tuttlepublishing.com

JAPAN
Tuttle Publishing, Yaekari Building, 3F, 5-4-12 Osaki, Shinagawa-ku, Tokyo 141-0032
Tel: (81) 3 5437-0171 | Fax: (81) 3 5437-0755
sales@tuttle.co.jp | www.tuttle.co.jp

ASIA PACIFIC
Berkeley Books Pte. Ltd., 3 Kallang Sector #04-01, Singapore 349278
Tel: (65) 6741-2178 | Fax: (65) 67414-2179
inquiries@periplus.com.sg | www.tuttlepublishing.com

First edition
25 24 23 22 21   11 10 9 8 7   2102TP
Printed in Singapore

Acknowledgments

I would like to acknowledge the origamists who have inspired my own work: Daniel Kwan, who has provided constant inspiration, and without whom this book wouldn't have occurred; the other young wireframe modular designers who have encouraged me to continue with my own work, including Aaron P. and Alec Sherwin; Meenakshi Mukerji, for offering advice about publishing; Francis Ow and Tomoko Fuse for the discovery of the first edge unit techniques applied in wireframe design; Tom Hull and Robert Lang through the inspiration their own works in wireframes have produced; Ekaterina Lukasheva, for advice about photography and inspiration in Kusudama/ decorative modular design. Thanks to Jeri McDaniel, for giving me the paper airplane calendar that got me started in origami, and to my parents for their constant help and support. Thanks to God, for helping me to strive for ever better work to glorify Him.

Books to Span the East and West

Our core mission at Tuttle Publishing is to create books which bring people together one page at a time. Tuttle was founded in 1832 in the small New England town of Rutland, Vermont (USA). Our fundamental values remain as strong today as they were then—to publish best-in-class books informing the English-speaking world about the countries and peoples of Asia. The world is a smaller place today and Asia's economic, cultural and political influence has expanded, yet the need for meaningful dialogue and information about this diverse region has never been greater. Since 1948, Tuttle has been a leader in publishing books on the cultures, arts, cuisines, languages and literatures of Asia. Our authors and photographers have won many awards and Tuttle has published thousands of titles on subjects ranging from martial arts to paper crafts. We welcome you to explore the wealth of information available on Asia at **www.tuttlepublishing.com.**